Advanced Pharmacology Glossary
&
Abbreviations

A Quick Reference Handbook
on Pharmacology

Advanced Pharmacology Glossary
&
Abbreviations

A Quick Reference Handbook
on Pharmacology

Dr. Pradeep K. Agarwal

Dr. Pradeep K. Agarwal

First Edition May 2013
New Delhi, 110047
Email: dr.pradeep.ka@gmail.com

The author and publisher have made conscientious effort to ensure that the information contained in this book is accurate and in accordance with accepted standards at the time of publication. However, in this rapidly changing world guidelines and practices are subject to change without any prior notification, therefore readers are advised to confirm these as and when needed.

ISBN-13: 978-1-499-62600-1

Book Contents

An Introduction to Pharmacology

Pharmacology is the branch of medicine and biology concerned with the study of drug action, where a drug can be broadly defined as any man-made, natural, or endogenous (within the body) molecule which exerts a biochemical and/or physiological effect on the cell, tissue, organ, or organism. More specifically, it is the study of the interactions that occur between a living organism and chemicals that affect normal or abnormal biochemical function.

The field encompasses drug composition and properties, synthesis and drug design, molecular and cellular mechanisms, organ/systems mechanisms, signal transduction/cellular communication, molecular diagnostics, interactions, toxicology, chemical biology, therapy, and medical applications and anti-pathogenic capabilities.

Sources of Drugs

The drugs are generally obtained from the following sources:

- **Minerals**: Liquid paraffin, magnesium sulfate, magnesium trisilicate, kaolin, etc.
- **Animals**: Insulin, thyroid extract, heparin and antitoxin sera, etc.
- **Plants**: Morphine, digoxin, atropine, castor oil, etc.
- **Synthetic source**: Aspirin, sulphonamides, paracetamol, zidovudine, etc.
- **Micro organisms**: Penicillin, streptomycin and many other antibiotics.
- **Genetic engineering:** Human insulin, human growth hormone etc.

Divisions of Pharmacology

Pharmacology is the basis of much of the research and development of new drugs. The future of pharmacology is not assured, as there remain many diseases for which cures nor has palliation been devised, for example, Alzheimer 's disease, AIDS, many forms of cancer. Even when a cure or

1

treatment is available, few medicines are perfect and the search for better drugs continues. The pharmacological sciences can be further subdivided:

- **Neuropharmacology** is the study of drugs on components of the nervous system, including the brain, spinal cord, and the nerves that communicate with all parts of the body. Neuropharmacologists study drug actions from a number of differing viewpoints.

- **Cardiovascular pharmacology** concerns the effects of drugs on the heart, the vascular system, and those parts of the nervous and endocrine systems that participate in regulating cardiovascular function.

- **Molecular pharmacology** deals with the biochemical and biophysical characteristics of interactions between drug molecules and those of the cell. It is molecular biology applied to pharmacological and toxicological questions.

- **Biochemical pharmacology** uses the methods of biochemistry, cell biology, and cell physiology to determine how drugs interact with, and influence, the chemical "machinery" of the organism.

- **Behavioural pharmacology** studies the effects of drugs on behaviour. Research includes topics such as the effects of psychoactive drugs on the phenomena of learning, memory, wakefulness, sleep, and drug addiction, and the behavioural consequences of experimental intervention in enzyme activity and brain neurotransmitter levels and metabolism.

- **Endocrine pharmacology** is the study of actions of drugs that are either hormones or hormone derivatives, or drugs that may modify the actions of normally secreted hormones.

- **Clinical pharmacology** is the application of pharmacodynamics and pharmacokinetics to patients with diseases and now has a significant pharmacogenetic component. Clinical pharmacologists study how drugs work, how they interact with the genome and with other drugs, how their effects can alter the disease process, and how disease can alter their effects.

- **Chemotherapy** is the area of pharmacology that deals with drugs used for the treatment of microbial infections and malignancies.

- **Systems and integrated pharmacology** is the study of complex systems and whole animal model approaches to best predict the efficacy and usefulness of new treatment modalities in human experiments. Results obtained at the molecular, cellular, or organ system levels are studied for their relevance to human disease through translation into research in whole animal systems.

- **Veterinary pharmacology** concerns the use of drugs for diseases and health problems unique to animals.

Routes of Administration

Route	Advantages	Disadvantages
EXTERNAL ROUTES		
Oral	Convenient, non sterile, good absorption of most drugs	Gastrointestinal irritation, potential for interactions, first pass destruction, inactivated by acids, variable absorption
Sublingual/ Buccal	Avoids first pass, avoids gastric acid	Few preparations suitable
Rectal	Avoids first pass, avoids gastric acid	Less dignified for the patient
PARENTERAL (refers to IV, IM and SC) ROUTES		
Intravenous (IV)	Rapid action, complete availability	Increased drug levels to heart, must be sterile, risk of sepsis and embolism
Intramuscular (IM)	Rapid absorption	Painful risk of tissue damage
Subcutaneous (SC)	Good for slower absorption	Absorption variable
Inhaled (Lungs)	Large absorption area, good for topical use	Few disadvantages
Other routes include intra-arterial, intrasternal, intrathecal, intra-articular, intraperitoneal, intraventricular, nasal, bronchial, vaginal, skin and conjunctiva		

Areas of Pharmacology

The two main areas of pharmacology are *pharmacodynamics and pharmacokinetics*. The former studies the effects of the drug on biological systems, and the latter the effects of biological systems on the drug.

Pharmacokinetics

It is defined as the study of the time course of drug absorption, distribution, metabolism, and excretion (ADME). Clinical pharmacokinetics is the application of pharmacokinetic principles to the safe and effective therapeutic management of drugs in an individual patient.

Pharmacodynamics

Pharmacodynamics refers to the relationship between drug concentration at the site of action and the resulting effect, including the time course and intensity of therapeutic and adverse effects. The effect of a drug present at the site of action is determined by that drug's binding with a receptor. Receptors may be present on neurons in the central nervous system (i.e., opiate receptors) to depress pain sensation, on cardiac muscle to affect the intensity of contraction, or even within bacteria to disrupt maintenance of the bacterial cell wall.

The majority of drugs are either a mimic or inhibit normal physiological/biochemical processes or inhibit pathological processes in animals or inhibit vital processes of endo or ecto-parasites and microbial organisms.

ADME

Absorption - the process of a substance entering the blood circulation.

Distribution - the dispersion or dissemination of substances throughout the fluids and tissues of the body.

Metabolism (or Biotransformation) - the irreversible transformation of parent compounds into daughter metabolites.

Excretion - the removal of the substances from the body. In rare cases, some drugs irreversibly accumulate in body tissue.

Elimination is the result of metabolism and excretion.

Pharmacokinetics describes how the body affects a specific drug after administration. Pharmacokinetic properties of drugs may be affected by elements such as the site of administration and the dose of administered drug. These may affect the absorption rate. A fifth process, Liberation has been highlighted as playing an important role in pharmacokinetics:

Liberation - the process of release of drug from the formulation. Hence LADME may sometimes be used in place of ADME in reference to the core aspects of pharmacokinetics.

Absorption

Before a drug can begin to exert any effect on the body it has to be absorbed into the body systems. This absorption process can be affected by many things but the main factor relating to absorption is the route of administration.

It is important that nurses understand the implications attached to choosing routes of administration of drugs based on their absorption. Many patients may need to have their medication administration tailored to their particular medical condition or the medication which they are prescribed, and this is an important factor to consider as it can impact on the patient's ability or desire to take their medication.

Other factors controlling the rate and reliability of drug absorption can be said to be physiological or physico-chemical. Physiological factors relate to human physiological functions:

- **Blood flow to absorbing site** - The better the blood supply to the area the greater the rate of absorption. Therefore if a person has a good circulation they will have the ability to absorb the drug well.

- **Total surface area for absorption** - The greater the surface area the greater the rate of absorption. The intestine has a very large surface area, making it an ideal target for drug absorption. This is why you will find that most drugs are given orally where possible.

- **Time of arrival and contact time at absorption site** - The longer the drug is in contact with the absorbing surface the greater the rate of absorption. This is why if a person is suffering from diarrhoea the chances of a drug given orally being absorbed completely are lowered and other means of administration must be considered.

Physico-chemical factors relate to the chemical make-up of the drug in relation to human physiological function:

- **Solubility**. How soluble is the drug in body fluids? As the body is made up of a large amount of water, drugs can dissolve readily. However, certain drugs do not dissolve into small enough particles to ensure rapid absorption.
- **Chemical stability**. Will it break down readily?
- **Lipid to water partition coefficient**. Is it more fat soluble than water soluble? This is an important area to consider. As your cells are made up of a phospho-lipid layer, any drug that can dissolve well in lipids will pass through your tissues far more rapidly. Examples of drugs that are highly lipid soluble are anaesthetic agents and benzodiazepines.

Degree of ionization. Some drugs are weak acids and weak bases (alkalis). These drugs tend to disassociate when given to a person.

This means that some of the drug remains active and some is inactive. Often this depends on the pH of the solution (i.e. its acidity or alkalinity) in which the drug is being dissolved. For example, a weak acid does not disassociate as much if dissolved in an acid environment. This means that the drug can cross membranes in a more active form than if it had been dissolved in a neutral or base solution.

Distribution

Once drugs have been administered and absorbed, they have to be distributed to their site of action. For some drugs that site is known and such drugs are available to give locally or topically. All other drugs need to be distributed throughout the body. There are four main elements to this:

- **Distribution into body fluids** - These are mainly plasma, interstitial fluid and intracellular fluid. Molecular targets for drugs are found in these areas.
- **Uptake into body tissues/organs** - Specific tissues take up some drugs – for example, iodine and thyroid gland.
- **Extent of plasma protein binding** - Plasma proteins such as albumin can bind drug molecules. This varies widely among drugs. Drugs bound to plasma proteins are pharmacologically inert; only free drugs are active. Some drugs do not bind (e.g. caffeine) and some are highly bound (e.g. warfarin which is 99 per cent bound to plasma proteins). Some drugs can displace others from their binding sites on the plasma proteins – for example, phenylbutazone can displace warfarin from plasma proteins. This is an important consideration for drugs which have this effect.
- **Passage through barriers** - The two main examples are the placenta and the blood-brain barrier (BBB). Drugs must be highly lipid soluble to pass across these barriers. If not, they may not be able to reach their site of action.

The factors which affect drug distribution are taken into consideration by drug companies when developing and formulating medications. While these factors are of interest, the nurse's role in monitoring drug distribution is mainly in monitoring the onset of the effect of, or the response to, the medication. If analgesia is given and the patient reports reduced or relieved pain, the drug has been distributed to its target site.

Biotransformation

Biotransformation of drugs is the process of metabolizing the parent drug compound and occurs mainly in the liver (hence the term hepatic metabolism) to different compounds called metabolites. The drug metabolite may have decreased, increased or undergone no change in pharmacological activity compared to the parent drug. It may also have a different activity. Some drugs are what are termed pro-drugs – that is the drug itself is pharmacologically inactive until it is metabolized by the liver to its active form. A good example is codeine, which is metabolized to morphine by the body. The metabolite is more polar (i.e. chemically charged) than the parent drug and therefore is more readily excreted by the kidney. Drug metabolism can influence dose and frequency of dosing. Drugs which are metabolized quickly have a short duration of action and need to be administered more often (two, three or four times daily). Drugs which are metabolized slowly can have a longer duration of action and may only need to be given on a once-daily basis.

Phase 1 Metabolism – Oxidation, Reduction, Hydrolysis;
Phase 2 Metabolisms – Conjugation.

Hepatic metabolism

Most drugs undergo phase I oxidation followed by phase II conjugation

Excretion

Once drugs have had their desired effect they need to be excreted by the body. Principles of excretion include renal elimination and clearance, secretion into bile for faecal elimination and enterohepatic recirculation. As previously outlined, some drug metabolites can also have pharmacological effects. If these compounds were not eliminated, they would accumulate in the bloodstream and could cause toxic and unwanted effects. The main method of renal elimination is by active glomerular filtration. This is where ionized drugs are actively secreted into the proximal tubule. These ionized compounds are actively excreted by the kidney and are 'pushed' out into urine. A more passive form of drug compound movement occurs in the distal tubule of the kidney. Here there is passive reabsorption and excretion of drug molecules and metabolites according to a concentration gradient. Molecules move from a high concentration to a lower concentration by diffusion. This applies to unionized compounds (drugs without charge), and prevents the entire dose of a drug being excreted at once. This helps to maintain circulating plasma levels to allow the drug effect to continue until the next dose is taken.

Excretion into bile is another method of eliminating drug molecules and metabolites. These are secreted from the liver into bile and into the gut for faecal elimination. As in renal excretion, not all of the drug and its metabolites are eliminated entirely at once. Some drugs undergo enterohepatic recirculation. This is where some of the drug is reabsorbed from the gut, back into the bloodstream and represented to the liver for further metabolism. This can help to maintain circulating levels of active molecules to prolong drug effect until the next dose. An important example of a drug that undergoes this is the combined oral contraceptive pill.

General and molecular aspects

It is important that nurses involved in medicines management are aware of the sites of action for many commonly used drugs. Drugs exert their effects at molecular (chemical) targets, of which there are many.

Receptors

The plasma membrane of a human cell is selectively permeable in that it helps control what moves in and out of the cell. The cell membrane consists of a thin structured bilayer of phospholipids and protein molecules. The surfaces of plasma membranes are generally studded with proteins that perform different functions, like the reception of nutrients. In biochemistry these protein molecules are referred to as receptors. Molecules which bind to these receptors are called ligands. Examples of ligands are neurotransmitters, hormones or drugs.

A large number of drugs, which are clinically effective, exert their action by interaction with receptors. Examples include:

- ligand-gated ion channels (ionotropic receptors) such as the GABAA receptor, which binds benzodiazepines;
- G-protein coupled receptors such as adrenoceptors;
- kinase-linked receptors such as the insulin receptor;
- nuclear receptors such as the thyroid receptor.

Ion channels

Ion channels provide receptors which drugs can interact with. Drug actions at ion channels can take two forms.

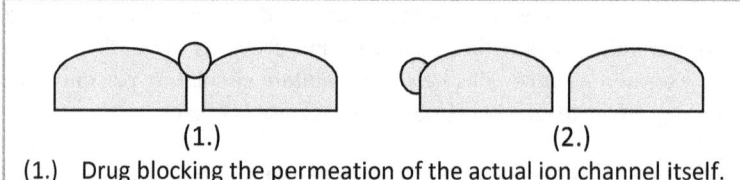

(1.)   (2.)

(1.) Drug blocking the permeation of the actual ion channel itself.
(2.) Drug binding the channel but not sitting within the channel

Drug binding the ion channel

The first form is known as channel blockers, whereby the drug blocks permeation of the channel, and the second are channel modulators whereby the drug binds to a receptor site within the ion channel and modulates permeation of the channel. This can happen by the drug altering the channel's response to its normal mediator.

Enzymes

Enzymes are biological catalysts that increase the rate of chemical reactions in the body. They are integral to many normal physiological functions. Many drugs target enzymes to prevent them from carrying out their normal function – for example, Enalopril acts on angiotensin converting enzymes, thereby preventing an increase in blood pressure.

Transport systems

These are also known as carrier molecule interactions. In some transmitter systems, there is normal physiological recycling of the transmitters, such as serotonin. After the release of serotonin from a neurone, it is taken back up by that same neurone using a serotonin-selective re-uptake system. The drug fluoxetine blocks the uptake transporter for serotonin as its mode of action. This results in an increased level of serotonin in the neuronal synapse. This mechanism has an onward effect which facilitates an increase in mood and makes fluoxetine and drugs similar to it good antidepressants.

Drug action

The time to the onset of drug action involves delivery of the drug to its site of action. This is largely controlled by:

- route of administration;
- rate of absorption;
- manner of distribution.

These are important considerations, as often we want the drug to have its effect within a certain time frame. We can speed up the time to the onset of drug action in many ways. If the drug is given orally, we can use liquid or dispersible formulations instead of regular tablets. If drug action is needed more quickly, we can use the intramuscular (IM) or intravenous (IV) route as necessary. For example, if a patient requires pain relief following myocardial infarction they would be given intravenous morphine rather than an oral preparation.

It is also possible to delay drug onset or prolong the effect by using enteric-coated or slow release preparations orally, or by using transdermal or subcutaneous (SC) routes. For example, people suffering with chronic pain from conditions such as rheumatoid arthritis may be given analgesia in the form of a transdermal patch. This is much preferred by the patient as it decreases the amount of oral analgesia required.

The duration of drug effect relates to the time it takes for the drug to be removed from its site of action. This is largely controlled by:

- rate of hepatic metabolism;
- rate of renal excretion.

It is important to be aware of the duration a drug will have its effect for. Drug companies do extensive studies to determine this information. They use the data they obtain to decide upon dosing schedules. It is vital that nurses know the normal dosing schedules for the drugs they are administering (this can easily be found in the British National Formulary – BNF) so that the correct regimen is implemented. Drugs need to be given more than once to have continued effect. Some drugs need to be given daily, while others need to be given two, three or four times per day to maintain effective action.

First pass metabolism

Some drugs undergo destruction by first pass metabolism. When absorbed through the stomach after oral administration, the drugs enter blood vessels which go directly to the liver. We call this the portal circulation. This means that drugs which are largely destroyed by liver enzyme systems will not enter the general systemic circulation. An example of such a drug is glyceryl trinitrate (GTN) which is metabolized completely by the liver at this stage. This is why you will find GTN being given via routes other than orally.

The concept of affinity

Drugs have what is termed an affinity for their receptors, or chemical targets. This is a measure of how well a drug can bind to its chemical target. The tighter the bond, the better the drug action. Some drugs have a higher affinity for their chemical targets than others. Those with a higher affinity will bind first, in preference to any other drug molecule present. Some drugs have a higher affinity for their targets than even the normal physiological molecule. This can be very useful in drug action, especially where the normal molecule is abundant and causing the problem or symptom the patient is experiencing. Higher affinity means that even small amounts of the drug will bind preferentially.

Agonistic and antagonistic drug action

Drugs can either be agonists or antagonists at their target sites. This is best explained using receptors as an example. When agonists or antagonists bind to receptors they are said to occupy the receptor site. The amount of drug occupying the receptor site relates to the magnitude of response to the drug itself. In simple terms the more of an agonist drug occupying a receptor, the greater the response. Agonists are drugs that bind to their targets and form a drug-receptor complex. Agonists activate the receptors to produce a response (known as full agonists) and have what is termed positive efficacy. Antagonists are drugs that bind to their targets and form a drug receptor complex, but without causing activation or response. They can block the receptor to its endogenous activator, thereby blocking normal function. They have what is termed zero efficacy. Receptor occupancy by antagonists is important if the drug is a competitive antagonist – i.e. it competes for occupancy with another drug or with the receptor's normal mediator. The amount of drug occupying will determine any response.

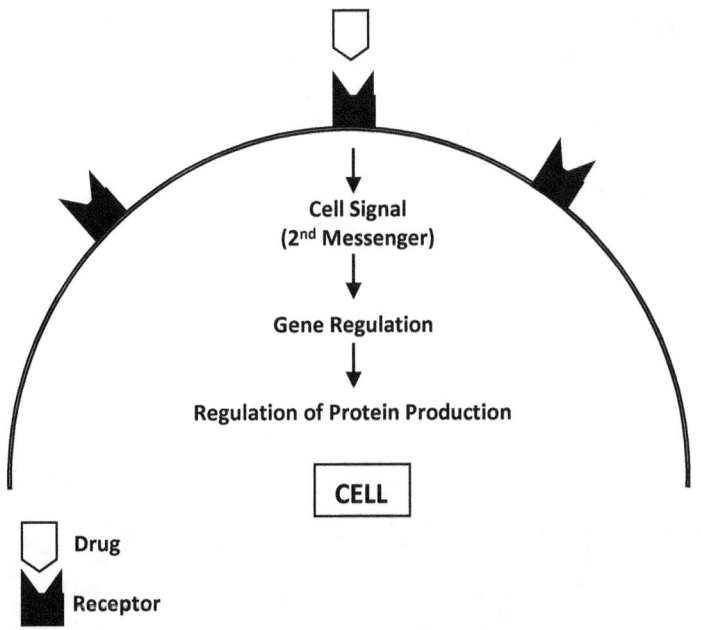

Cell Signal
(2nd Messenger)

Gene Regulation

Regulation of Protein Production

CELL

Drug

Receptor

This is a simplistic view of the concepts of agonism and antagonism as the response of a drug at its chemical target is actually graded.

For agonists we have:

Partial agonists: drugs that bind to their targets and activate them to produce a response which is less than that we would expect from a full agonist. They have what is termed partial efficacy.

Inverse agonists: drugs that bind to their targets and can reduce the normal activity of that chemical target. They have what is termed negative efficacy.

For antagonists we have:

Competitive antagonists: drugs that bind to the chemical targets and prevent activation by the normal target agent.

Non-competitive antagonists: drugs that do not necessarily bind to the chemical target but at a point in the chain of events block target activation.

Drug Specificity

Very few drugs are specific for their intended targets within the body. A prescriber will give a drug with a specific action in mind, for example salbutamol. Salbutamol is a beta2 adrenoceptor agonist. This means it has its main action at beta2 adrenoceptors in the bronchi. This gives us its desired effect as a bronchodilator which eases breathing in asthma. However the action of salbutamol is not that specific and can act on other beta2 adrenoceptors in the body as well as on beta adrenoceptors, especially if given in higher doses leading to increased receptor occupancy. This is the reason that some of the side-effects of drugs can be seen. In the case of salbutamol, action at other beta adrenoceptors can lead to palpitations and increased occupancy at non-bronchial beta adrenoceptors can cause tremor.

A

Glossary

1. **Absolute risk**
   Risk in a population of exposed persons; the probability of an event affecting members of a particular population (e.g. 1 in 1,000). Absolute risk can be measured over time (*incidence*) or at a given time (prevalence).

2. **Absorbance**
   Absorbance is used for assays such as ELISA assays, protein and nucleic acid quantification or enzyme activity assays (i.e. in the MTT assay for cell viability). A light source illuminates the sample using a specific wavelength (selected by an optical filter, or a monochromator), and a light detector located on the other side of the well measures how much of the initial (100 %) light is transmitted through the sample: the amount of transmitted light will typically be related to the concentration of the molecule of interest.

3. **Absorption Rate Constant**
   The rate at which a medication is absorbed from dosage site to measurement location. This is applicable to all drugs except intravenous medications.

4. **Acute**
   Drug to improve a life threatening condition. (ex: epinephrine for anaphylaxis).

5. **Accumulation, Accumulation Ratio**
   The amount of a medication found within a bodily fluid at a specific point when a steady state has been attained. The point of equality between drug administration and drug elimination.

13

6. **Accuracy**
The amount of error found in the results of a scientific equation.

7. **Activity – intrinsic**
The property of a drug which determines the amount of biological effect produced per unit of drug-receptor complex formed. Two agents combining with equivalent sets of receptors may not produce equal degrees of effect even if both agents are given in maximally effective doses; the agents differ in their intrinsic activities and the one producing the greater maximum effect has the greater intrinsic activity. Intrinsic activity is not the same as "potency" and may be completely independent of it. Meperidine and morphine presumably combine with the same receptors to produce analgesia, but regardless of dose, the maximum degree of analgesia produced by morphine is greater than that produced by meperidine; morphine has the greater intrinsic activity. Intrinsic activity - like affinity - depends on the chemical natures of both the drug and the receptor, but intrinsic activity and affinity apparently can vary independently with changes in the drug molecule

8. **Acquired Disease**
Any disease triggered by external factors and not directly caused by a person's genes.

9. **Addiction**
A situation where use of a drug has changed the behavior and methods of the user, creating a need for it in order to continuing using or to obtain more of it.

10. **Additive Effects**
Drug interactions in which the effect of a combination of two or more drugs with similar actions is equivalent to the sum of the individual effects of the same drugs given alone (1+1=2) compare with synergistic effects.

11. **Adherence**
Implementation or fulfillment of a prescriber's or caregiver's prescribed course of treatment or therapeutic plan by a patient. (Similar to compliance).

12. **ADME**
An acronym in pharmacokinetics and pharmacology for absorption, distribution, metabolism, and excretion, and describes the disposition of a pharmaceutical compound within an organism.

13. Adrenergic
Pertaining to the neurotransmitter norepinephrine.

14. Adverse Drug Event
Any undesirable occurrence related to administering or failing to administer a prescribed medication.

15. Adverse Drug Reaction
Any unexpected, unintended, undesired, or excessive response to a medication given at therapeutic dosages (as opposed to overdose).

16. Adverse Effects
A general term for any undesirable effects that are a direct response to one or more drugs.

17. Affect
The predominant emotion in a person's mental state.

18. Affective Domain
The most intangible component of the learning process. Affective behavior is conduct that expresses feelings, needs, beliefs, values, and opinions.

19. Affinity (drug)
The equilibrium constant of the reversible reaction of a drug with a receptor to form a drug-receptor complex; the reciprocal of the dissociation constant of a drug-receptor complex. Under the most general conditions, where there is a 1:1 binding interaction, at equilibrium the number of receptors engaged by a drug at a given drug concentration is directly proportional to their affinity for each other and inversely related to the tendency of the drug-receptor complex to dissociate. Obviously, affinity depends on the chemical natures of both the drug and the receptor. "Affinity" is not the same as "duration of action".

20. Aging
Inhibition of acetycholinesterase (AchE) with organophosphates results in a increase in Ach levels. If allowed to associate with AchE for certain period of time a phenomenon called 'aging' occurs, involving the loss of a group attached to phosphorus and leading to the formation of a negatively charged irreversibly phosphorylated AchE enzyme. The aging process can be very short (ie. nerve gases, secs) or longer (ie. pesticides, hrs). Pralidoxime (2-PAM) can regenerate AchE from the organophosphate but only before the 'aging' process.

21. **Agonist**

A drug that binds to and activates a receptor. Can be *full*, *partial* or *inverse*. A *full agonist* has high efficacy, producing a full response while occupying a relatively low proportion of receptors. A *partial agonist* has lower efficacy than a full agonist. It produces sub-maximal activation even when occupying the total receptor population, therefore cannot produce the maximal response, irrespective of the concentration applied. An *inverse agonist* produces an effect opposite to that of an agonist, yet binds to the same receptor binding-site as an agonist.

   a. **Agonist, Partial**: A partial agonist is an agonist that produces a maximal response that is less than the maximal response produced by another agonist acting at the same receptors on the same tissue, as a result of lower intrinsic activity.

   b. **Agonist, Full**: A full agonist is an agonist that produces the largest maximal response of any known agonist that acts on the same receptor.

   c. **Agonist, Inverse**: An inverse agonist is a ligand that by binding to a receptor reduces the fraction of receptors in an active conformation, thereby reducing basal activity. This can occur if some of the receptors are in the active form in the absence of a conventional agonist.

22. **Agranulocytosis**

A disorder in which there is severe deficiency of certain blood cells (leucocytes) as a result of damage to the bone marrow by toxic drugs or chemicals. It is characterized by fever, with ulceration of the mouth and throat, and can lead rapidly to death.

23. **Akathisia**

A condition of motor restlessness in which there is a feeling of muscular quivering, an urge to move about constantly, and an inability to sit still often exhibited as pacing or rocking. Often accompanied by sensations of muscular discomfort, dysphoria and agitation.

24. **Allergic Response**

A situation in which the body forms antibodies against a specific drug, causing a physical reaction that may or may not be severe.

25. **Alleles**

The two or more alternative forms of a gene that can occupy a specific locus (location) on a chromosome.

26. Allergic Reaction

An immunologic hypersensitivity reaction resulting from the unusual sensitivity of a patient to a particular medication; a type of adverse drug event.

27. Alternative Medicine

Herbal medicine, chiropractic, acupuncture, reflexology, and any other therapies traditionally not emphasized in Western medical schools but popular with many patients.

28. Allopathy

Non-traditional, western scientific therapy, usually using synthesised ingredients, but may also contain a purified active ingredient extracted from a plant or other natural source; usually in opposition to the disease.

29. Allosteric Modulator

A drug that binds to a receptor at a site distinct from the active site. Induces a conformational change in the receptor, which alters the affinity of the receptor for the endogenous ligand. Positive allosteric modulators increase the affinity, whilst *negative allosteric modulators* decrease the affinity.

30. Amoxicillin

Medicine to treat bacterial infections (ex: Penicillin).

31. Amplification:

The quantity of change in determined output per unit change in input.

32. Analgesic

A medication that alleviates pain without the patient losing consciousness.

33. Anaphylaxis

Serious and rapid allergic reaction usually involving more than one part of the body which, if severe enough, can be fatal. Usually associated with bee or wasp stings but is more common with food or drug allergies. **Treatment:** Epinephrine (im) is the drug of choice.

34. Analyses of secular trends

Examine trends in disease events over time or across different geographical locations and correlate them with trends in putative exposures, such as rates of drug utilization. The unit of observation is a subgroup of a population rather than individuals.

35. Analytic studies
Studies with control groups, namely case-control studies, cohort studies, and randomized clinical trials.

36. Analysis of variance (ANOVA)
A statistical analysis involving the comparison of variance reflecting different sources of variability.

37. Anesthetic
A drug that causes loss of sensation. *General anesthetics* cause not only loss of sensation, but also loss of consciousness

38. Animal source
Glandular substance (ex: Insulin; Thyroid hormone).

39. Anesthetic
A medication that causes loss of sensation. This is sometimes used to alleviate pain or for loss of consciousness for surgical procedures.

40. Antacid
Medicine to decrease stomach acid (ex: Nexium).

41. Antagonism
The effect of two or more drugs such that the combined effect is less than the sum of the effects produced by each agent separately. The agonist is the agent producing the effect which is diminished by the administration of the antagonist. Antagonisms may be any of three general types:

Chemical: caused by combination of agonist with antagonist, with resulting inactivation of the agonist

Physiological: caused by agonist and antagonist acting at two independent sites and inducing independent, but opposite effect.

Pharmacological: caused by action of the agonist and antagonist at the same site i.e. epinephrine and propranolol at beta-receptors

42. Antagonist
A drug that attenuates the effect of an agonist. Can be competitive or non-competitive, each of which can be reversible or irreversible. A competitive antagonist binds to the same site as the agonist but does not activate it, thus blocks the agonist's action. A non-competitive antagonist binds to an allosteric (non-agonist) site on the receptor to prevent activation of the receptor. A reversible antagonist binds non-covalently to the receptor, therefore can be "washed out". An irreversible antagonist binds covalently

to the receptor and cannot be displaced by either competing ligands or washing.

43. **Anticipated harmful effects**
Unwanted effects of drugs that could have been predicted on the basis of existing knowledge.

44. **Antidiuretic**
Medicine to decrease urinary output (to treat dehydration).

45. **Antiemetic**
Medicine to treat / prevent nausea and vomiting (emisis = vomiting) (ex: Phenergan).

46. **Antihypertensive**
Medicine to treat high blood pressure.

47. **Antipyretic**
Fever reducer

48. **Antitussive**
Medicine to treat / prevent coughing (ex: Robitussin)

49. **Antiviral**
Medicine to treat viruses.

50. **Anxiolytic**
Anti-anxiety drug.

51. **Area Under the Curve**
The area on a graph that falls under the curve when plotting time after administration of a drug against the plasma concentration of a drug. It is used to estimate how long it takes for a drug to be removed from the body.

52. **Association**
Events associated in time but not necessarily linked as cause and effect.

53. **Attributable risk**
Difference between the risk in an exposed population (*absolute risk*) and the risk in an unexposed population (*reference risk*). Attributable risk is the result of an absolute comparison between outcome frequency measurements, such as incidence.

54. Autonomic ganglia
Groups of autonomic nerve cells located outside the central nervous system.

55. Autonomic nervous system
Innervations of smooth muscle, glands and visceral organs, which are not normally under voluntary control. Subdivided principally into the sympathetic and parasympathetic efferent systems. Autonomic reflexes are reflexes that act through these efferent systems; their afferent pathways may be either the same as pathways that sub-serve conscious perceptions (as with salivation) or they may be different (as with baroreceptor reflexes). The afferent pathways are not distinctive in any anatomical way, and are not usually described as 'autonomic' except by association with particular reflex actions

56. Availability
This is the amount of a drug dosage that is absorbed into circulation after administration of a specific dosage. (Also referred as bioavailability).

Abbreviations

1.	a	Before
2.	aa	Of each
3.	AAA	Abdominal aortic aneurysm
4.	abd./abdo.	Abdomen
5.	AB./abort.	Abortion
6.	ABG	Arterial blood gases
7.	Abn	Abnormal
8.	Ac	Before meals
9.	A D	Right ear
10.	ADH	Anti-diuretic hormone
11.	ad lib	As desired
12.	ADM	Admission
13.	AF	Atrial fibrillation
14.	AFP	Alpha feto protein
15.	AGN	Acute glomerulonephritis
16.	AIDS	Acquired immune deficiency syndrome
17.	AL	Antero-lateral

18.	Alb	Albumin
19.	ALS	Amyotrophic lateral sclerosis
20.	AMA	Against medical advice or American Medical Association
21.	AMS	Altered mental status
22.	alt. dieb.	On alternate days
23.	AM, am	Morning
24.	Amp	Ampule
25.	amt.	Amount
26.	Ante	Before
27.	aq.	Water
28.	A S	Left ear
29.	A S A	Acetylsalicylic acid
30.	ASAP	As soon as possible
31.	ATC	Around the clock
32.	A U	Both ears

B

Glossary

1. **Bacterial / Fungi**
   Simple organisms produce substances to make antibiotics: (ex: PCN - penicillin; cephalosporins).

2. **BALB/c**
   An albino strain of laboratory mouse from which a number of common sub-strains are derived. BALB/c sub-strains are "particularly well known for the production of plasmacytomas on injection with mineral oil," an important process for the production of monoclonal antibodies. They are also reported as having a "low mammary tumor incidence, but do develop other types of cancers in later life, most commonly reticular neoplasms, lung tumors, and renal tumors.

3. **Baroreceptor reflex**
   Baroreceptors found in the aorta arch and carotid sinuses, sense changes in blood pressure. As blood pressure goes up, the baroreceptors are stimulated and they deliver a higher rate of impulses to the vasomotor center of the brain. This causes a reduction in sympathetic tone and a stimulation of vagal tone. As a result, there is a reduction in heart rate, cardiac contractility, and vasodilation of blood vessels throughout the body which all contribute to lower blood pressure. If blood pressure goes down, baroreceptors reduce their rate of firing, causing the opposite effect. The baroreceptor reflex is more sensitive to rapidly changing pressure (standing up, or sitting down) than to a constantly elevated or depressed pressure. Baroreceptors will adapt to long term increased or decreased blood pressure.

4. **Before and after study**
   A situation in which the investigator compares outcomes before and after the introduction of a new intervention.

5. **Belladonna alkaloids**
   Group of alkaloids, including atropine and scopolamine, found in plants such as belladonna and jimsonweed. They are used in medicine to dilate the pupils of the eyes, dry respiratory passages, prevent motion sickness, and relieve cramping of the intestines and bladder.

6. **Benefit**
   An estimated gain for an individual or a population.

7. **Benefit - risk analysis**
   Examination of the favorable (beneficial) and unfavorable results of undertaking a specific course of action. (While this phrase is still commonly used, the more logical pairings of benefit-harm and effectiveness-risk are slowly replacing it).

8. **Benign prostrate hypertrophy (hyperplasia)**
   It is an enlargement of the prostate gland. This can often compress the urethra and partially block urine flow. Prostate enlargement adversely affects about half the men in their 60s and close to 80 percent of men in their 80s. The presence or absence of prostate gland enlargement is not related to the development of prostate cancer. Treatment: Alpha1 blockers such as prazosin or terazosin (Hytrin).

9. **Bias**
   Any systematic error in a measurement process. One common effort to avoid bias in research studies involves the use of blinded study designs.

10. **Bioassay or Biological Assay**
    Establishing the strength of a chemical, physical, or biological agent, by way of a biological marker.

11. **Bioavailability**
    The percent of dose entering the systemic circulation after administration of a given dosage form. More explicitly, the ratio of the amount of drug "absorbed" from a test formulation to the amount "absorbed" after administration of a standard formulation. Frequently, the "standard formulation" used in assessing bioavailability is the aqueous solution of the drug, given intravenously.

12. Biopharmaceutics
The study of how the pharmaceutical expression of certain drugs can impact their pharmacodynamic and pharmacokinetic behavior.

13. Biological products
Medical products prepared from biological material of human, animal or microbiologic origin (such as blood products, vaccines, insulin).

14. Biotransformation
The chemical change of a drug that happens due to the effects the body has on it.

15. Biotranslocation
The transfer and movement of drugs in and throughout biological organisms.

16. Black Box Warning
A type of warning that appears in a drug's prescribing information, required by the US FDA alerting prescribers of serious adverse events that have occurred with the given drug

17. Blind Experiment
A type of experiment in which the participants are unaware of the drug doses or treatments involved, so as not to affect the outcome.

18. Blinded Investigational Drug Study
A research design in which the subjects are purposely unaware of whether the substance they are administered is the drug under study or a placebo. This method serves to eliminate bias on the part of research subjects in reporting their body's responses to investigational drugs.

19. Blood-Brain Barrier
The barrier system that restricts the passage of various chemicals and microscopic entities between the bloodstream and the central nervous system. It still allows for the passage of essential substances such as oxygen.

20. Bolus
A single often times large dose of medication.

21. Bonferroni Correction
This is a multiple comparison technique used to adjust an error level (p value) to allow for multiple tests.

22. Bradykinesia
A symptom of parkinsonism comprising a difficulty in initiating movements and slowness in executing movements and maintaining body posture.

Abbreviations

1.	B	Body weight. Sometimes, as a subscript, to indicate "of, or in, the body"; thus, A_B is the amount of drug in the body.
2.	B	The slope of a linear plot of log C against t, when logarithms to the base 10, common logarithms, are used; the slope of the linear, semi-logarithmic, plot of a first-order reaction when common logarithms are used. $k_{el} = 2.303b$; $t_{1/2} = 0.301/b$.
3.	b a	barium
4.	baso.	Basophil
5.	BBB	Bundle branch block
6.	BBx	Bone biopsy
7.	BE	Barium enema; base excess
8.	Bid	Twice a day
9.	bilat.	Bilateral
10.	bili.	Bilirubin
11.	Bin	Two times a night
12.	bis i. d.	Twice a day
13.	Bld	Blood
14.	BM	Bowel movement, bone marrow
15.	B_{max}	The maximum amount of drug or radioligand, usually expressed as picomoles (pM) per mg protein, which can bind specifically to the receptors in a membrane preparation. It can be used to measure the density of the receptor site in a particular preparation.
16.	BOM	Bilateral otitis media
17.	Bo	On a graph, the slope that occurs when concentration is plotted against the drug half life (or C is plotted against t).

18.	Bp	Blood pressure
19.	BPH	Benign prostatic hypertrophy
20.	BS	Bowel sounds, breath sounds
21.	BSA	Body surface area
22.	BSO	Bilateral salphingoooophorectomy
23.	BTL	Bilateral tubal ligation
24.	BUN	Blood urea nitrogen
25.	Bx	Biopsy

C

Glossary

1. **Case control study**

   An epidemiological study involving the observation of cases (persons with the condition) and a suitable control (comparison, reference) group of persons without the condition. The relationship of an attribute to the condition is examined by comparing retrospectively the past history of the people in the two groups with regard to how frequently the attribute is present.

2. **Case series**

   A descriptive study of a subset of a defined population (i.e., a single patient or group of patients) which aims to describe the association between factors or attributes which the sample is exposed to, and the probability of occurrence of a given disease or other outcome. Case series are collections of individual case reports, which may occur within a fairly short period of time.

3. **Causality**

   The relating of cause to the effect produced. A cause is termed "necessary" when the variable must always precede the event; "sufficient" if the variable inevitably initiates or produces the effect. Any given cause may be necessary, sufficient, neither or both.

4. **Causality assessment**

   The evaluation of the likelihood that a medicine was the causative agent of an observed adverse reaction. Causality assessment is usually made according established algorithms.

5. **Caveat document**

   The formal advisory warning accompanying data release from the WHO

Global ICSR Database: it specifies the conditions and reservations applying to interpretations and use of the data.

6. **Ceiling (drug)**
The maximum biological effect that can be induced in a tissue by a given drug, regardless of how large a dose is administered. The maximum effect produced by a given drug may be less than the maximum response of which the reacting tissue is capable, and less than the maximum response which can be induced by another drug of greater intrinsic activity. "Ceiling" is analogous to the maximum reaction velocity of an enzymatic reaction when the enzyme is saturated with substrate.

7. **CemFlow**
Software developed by UMC for collection and analysis of data in Cohort Event Monitoring.

8. **Certain drug causality**
A clinical event, including laboratory test abnormalities, occurring in a plausible time relationship to drug administration, and which cannot be explained by concurrent disease or other drugs or chemicals. The response to withdrawal of the drug (dechallenge) should be clinically plausible. The event must be definitive pharmacologically or phenomenologically, using a satisfactory rechallenge procedure if necessary.

9. **Chemical Name**
The name that describes the chemical composition and molecular structure of a drug/

10. **Chemotherapy**
The treatment of cancerous or parasitic illnesses, where the drug affects only the neoplastic cells or invading organisms.

11. **Cheng-Prusoff Equation**
Used to determine the K_i value from an IC_{50} value measured in a competition radioligand binding assay:

$$K_i = \frac{I\, C_{50}}{1 + \frac{[S]}{K_m}}$$

Where [L] is the concentration of free radioligand, and K_d is the dissociation constant of the radioligand for the receptor.

12. Cholinergic
Pertaining to the neurotransmitter acetylcholine.

13. Chromatin
A collective term for all of the chromosomal material within a given cell.

14. Chromosomes
Structures in the nuclei of cells that contain linear threads of deoxyribonucleic acid (DNA), which transmits genetic information, and are associated with ribonucleic acid (RNA) molecules and synthesis of protein molecules.

15. Chronotrope
Drug or chemical that affects the heart rate.

16. Clearance
Clearance of a chemical is the volume of body fluid from which the chemical is, apparently, completely removed by biotransformation and/or excretion, per unit time. In fact, the chemical is only partially removed from each unit volume of the total volume in which it is dissolved. Since the concentration of the chemical in its volume of distribution is most commonly sampled by analysis of blood or plasma, clearances are most commonly described as the "plasma clearance" or "blood clearance" of a substance.

17. Clinical pharmacology
The study of the effects of drugs in humans.

18. Clinical Therapeutic Index
An assessment of a drug having more safety at an acceptable level of potency or more potency at an acceptable level of safety within the recommended drug dosage.

19. Clustering
A closely grouped series of events or cases with well defined distribution patterns, in relation to time or place or both.

20. Coenzyme
A non-protein organic compound that in the presence of an enzyme, plays an essential role in the reaction that is catalysed by the enzyme.

21. Cohort study
The analytic method of epidemiological study in which subsets of a defined population can be identified who are, have been, or in the future may be exposed or not exposed in different degrees, to a factor or factors hypothesised to influence the probability of occurrence of a given disease or other outcome. Studies usually involve the observation of a large population, for a prolonged period (years).

22. Confidence interval
The computed interval with a given probability – e.g., 95%, that the true value of a variable such as a mean, proportion, or rate is contained within the interval. The 95% CI is the range of values in which it is 95% certain that the true value lies for the whole population.

23. Confounder
A third variable that indirectly distorts the relationship between two other variables, because it is independently associated with each of the variables.

24. Confounding
A situation in which the measure of the effect of an exposure on risk is distorted because of the association of exposure with other factor(s) that influence the outcome under study.

25. Cognitive Domain
The domain involved in the learning and storage of basic knowledge. It is the thinking portion of the learning process and incorporates a person's previous experiences and perceptions.

26. Cohort Event Monitoring
Cohort Event Monitoring (CEM) is a prospective, observational study of events that occur during the use of medicines, for intensified follow-up of selected medicinal products phase. Patients are monitored from the time they begin treatment, and for a defined period of time.

27. Compartment(s)
The area within the body that a drug tends to dwell in after it has been absorbed.

28. Competitive binding
It describes the event whereby there are two different molecules which have very similar but not identical chemical structures. Both structures are complementary to, and will have affinity to bind to, an effector such as a

receptor or enzyme, however normally only one will be able to produce a response. This situation can be considered using competitive antagonism as an example. A competitive antagonist, if present, will compete with agonist to bind to a receptor, without activating it – but in such a way to prevent the binding of the agonist. That is, the agonist and antagonist are in competition to bind to a receptor since the receptor can only bind to one of them at a time. At any agonist concentration, there will be reduced receptor occupancy by agonist in the presence of antagonist. However, because the two are in competition, raising the concentration of agonist will allow agonist occupancy to increase. This form of antagonism is surmountable and maximum response and occupancy can be achieved by raising the agonist concentration enough.

29. Complementary Medicine
Alternative medicine when used simultaneously with, rather than instead of, standard Western medicine (using conventional medicine and alternative medicine at the same time).

30. Compliance
The level of cooperation of a patient when following a prescribed treatment regimen.

31. Contraindication
Any condition, especially one related to a disease state or other patient characteristic, including current or recent drug therapy, that renders a particular form of treatment improper or undesirable.

32. Control group
The comparison group in drug-trials not being given the studied drug.

33. Controlled Substances
Any drugs listed on one of the "schedules" of the Controlled Substance Act (also called scheduled drugs).

34. Conventional Medicine
The practice of medicine as taught in Western medical schools.

35. Coombs Test
It is used to detect autoantibodies against your own red blood cells (RBCs). Many diseases and drugs (e.g., quinidine, methyldopa and procainamide) can lead to production of these antibodies. The test is only rarely used to diagnose a medical condition but is essential for use by laboratories such as blood banks. Blood banks use the Coombs' test is to determine whether

there is likely to be an adverse reaction to blood that is going to be used for a blood transfusion.

36. Cross-over experiment
A form of experiment in which each subject receives the test preparation at least once, and every test preparation is administered to every subject. At successive experimental sessions each preparation is "crossed-over" from one subject to another. The purpose of the cross-over experiment is to permit the effects of every preparation to be studied in every subject, and to permit the data for each preparation to be similarly and equally affected by the peculiarities of each subject.

37. Critical terms
Some of the terms in WHO-ART are marked as 'Critical Terms'. These terms either refer to or might be indicative of serious disease states, and warrant special attention, because of their possible association with the risk of serious illness which may lead to more decisive action than reports on other terms.

38. Cross-Tolerance
Tolerance to a drug that generalizes to drugs that are chemically related of that produce similar affects. For example, a patient who is tolerant to heroin will also exhibit cross-tolerance to morphine.

39. CT Index
The measure of the effects of a drug as calculated by plotting drug concentration against time.

40. Cumulative action
It occurs when a drug is administered in several doses, causing an increased effect. This is due to a quantitative buildup of the drug in the blood.

41. Cycloplegia
Paralysis or loss of function of the ciliary muscle; this results in loss of accommodation (ability to focus).

Abbreviations

1.	C_{ss}	The concentration of a drug or chemical in a body fluid – usually plasma – at the time a "steady state" has been achieved, and rates of drug administration and drug elimination are equal. C_{ss} is a value approached as a limit and is achieved, theoretically, following the last of an infinite number of equal doses given at equal intervals.
2.	C57BL/6	C57BL/6 often referred to as "C57 black 6" or just "black 6" is a common inbred strain of lab mouse. Dark brown, nearly black, coat. Easily irritable temperament. They have a tendency to bite. The immune response of mice from the C57BL/6 strain distinguish it from other inbred strains like BALB/c.
3.	C, C_x	The concentration (in units of mass/volume) of a chemical in a body fluid such as blood, plasma, serum, urine, etc.; the specific fluid may be indicated by a subscript, i.e. C_u, the concentration of drug in the urine; when no subscript is used, C is commonly taken to be the concentration in the plasma.
4.	c	About, approximately
5.	CA	Carcinoma/ cancer
6.	Ca	Calcium
7.	CABG	Coronary artery bypass graft
8.	CAC	Cardiac arrest code
9.	CAD	Cornary artery disease
10.	cal.	Calorie
11.	c with line above it	With
12.	cap	Capsule
13.	CAT (scan)	Computerized axial tomography
14.	C capitalized	Cup, Celsius
15.	CBC	Complete blood count
16.	C C capitalized	Chief complaint
17.	c forward slash o	Complaining of
18.	cc	Cubic centimeter

19.	CD	Controlled dose
20.	CHB	Complete heart block
21.	CHD	Coronary heart disease
22.	CHF	Congestive heart failure
23.	chol.	Cholesterol
24.	CIN	Cervical intraepithelial neoplasia
25.	CIS	Carcinoma in situ
26.	Cl	Chloride
27.	CLL	Chronic lymphocytic leukemia
28.	Cm	Centimeter
29.	C_{max}	The maximum or "peak" concentration (C_{max}) of a drug observed after its administration; the minimum or "trough" concentration (C_{min}) of a drug observed after its administration and just prior to the administration of a subsequent dose. For drugs eliminated by first-order kinetics from a single-compartment system, C_{max}, after n equal doses given at equal intervals is given by $C_0(1 - f^n)/(1 - f) = C_{max}$, and $C_{min} = C_{max} - C_0$.
30.	C_{min}	The time following drug administration at which the peak concentration of C_{max} occurs, t_p (for any route of administration but the intravenous), is given by $t_p = (\ln k_a - \ln k_{el})/(k_a - k_{el})$. (Remember that ln is the natural logarithm, to the base e, rather than the common logarithm or logarithm to the base 10; ln X=2.303 log X.).
31.	CMV	Cytomegalovirus
32.	CNS/cns	Central nervous system
33.	C_0	The fictive concentration of a drug or chemical in the plasma at the time (in theory) of an instantaneous intravenous injection of a drug that is instantaneously distributed to its volume of distribution. C_0 is determined by extrapolating, to zero-time, the plot of log C against t (for apparently "first-order " decline of C) or of C against t (for apparently "zero-order" decline of C).
34.	C/O	Cardiac output
35.	CO_2	Carbon dioxide
36.	Comp	Compound

37.	cont.	Continuous or continue
38.	cont. rem.	Let the remedies be continued
39.	COR	Heart
40.	CP	Chest pain
41.	CPA	Costrophrenic angle
42.	CPAP	Continuous positive airway pressure
43.	CPK	Cardiopulmonary resuscitation
44.	CPR	Cardiopulmonary resuscitation
45.	CR	Controlled release
46.	CRF	Chronic renal failure
47.	C-section	Caesarean section
48.	C&S	Culture and sensitivity
49.	CSF	Cerbrospinal fluid
50.	CTX	Contractions
51.	CV	Cardiovascular
52.	CVA	Cerebrovascular accident
53.	CVAT	Costovertebral disease
54.	CVP	Central venous pressure
55.	Cx	Cervix
56.	CXR	Chest X-ray
57.	CYP	The cytochrome P450 superfamily. The function of most CYP enzymes is to catalyze the oxidation of organic substances. The most common reaction catalyzed by cytochromes P450 is a monooxygenase reaction. RH (organic substrate) + O_2 + 2H+ + 2e− → ROH + H_2O. CYP families in humans divided among 18 families of cytochrome P450 genes and 43 subfamilies.

Glossary

1. **Data mining**
   A general term for computerized extraction of potentially interesting patterns from large data sets, often based on statistical algorithms. A related term with essentially the same meaning is 'pattern discovery'. In pharmacovigilance, the commonest application of data mining is so called disproportionality analysis, for example using the Information component (IC).

2. **Dechallenge**
   The withdrawal of a drug from a patient; the point at which the continuity, reduction or disappearance of adverse effects may be observed.

3. **Decongestant**
   A drug that relieves congestion, e.g. pseudoephedrine.

4. **Demulcent**
   A soothing medication used to relieve pain in inflamed tissues.

5. **Deoxycorticosterone**
   A steroid hormone, produced in the adrenal cortex; sometimes manufactured synthetically for use in cases of adrenal insufficiency.

6. **Dependence**
   A physical need to maintain administration of a specific drug in order to avoid withdrawal symptoms.

7. **Depressant**
   A medication that decreases or lessons a body function or activity.

8. **Desensitization**

A reduction in response to an agonist while it is continuously present at the receptor, or progressive decrease in response upon repeated exposure to an agonist.

9. **Desipramine**

A tricyclic, heterocyclic drug used to treat depression.

10. **Diacetylmorphine**

Chemical name for heroine.

11. **Digestive**

A substance that aids digestion.

12. **Digitalis**

A genus of herbaceous shrubs of the *Scrophulariaceae* family, including the foxglove, *Digitalis purpurea*.

13. **Disintegration Time**

The time it takes for a drug tablet to dissolve into pieces of a set size or smaller.

14. **Dispense**

To issue, distribute, or put out.

15. **Disproportionality analysis**

Screening of ICSR databases for reporting rates which are higher than expected. For drug-ADR pairs, common measures of disproportionality are the Proportional Reporting Ratio (PRR), the Reporting Odds Ratio (ROR), The Information Component (IC), and the Empirical Bayes Geometrical Mean (EBGM). There are also disproportionality measures for drug-drug-ADR triplets, such as Omega (Ω).

16. **Dissolution time**

The time required for a given amount (or fraction) of drug to be released into solution from a solid dosage form. Dissolution time is measured in vitro, under conditions which simulate those which occur in vivo, in experiments in which the amount of drug in solution is determined as a function of time. Needless to say, the availability of a drug in solution - rather than as part of insoluble particulate matter - is a necessary preliminary to the drug's absorption.

17. Distribution

The volume within a person in which the administrated drug appears to have been dispersed. Also known as volume of distribution.

18. Diuretic

Medicine to increase urinary output, decrease blood pressure (to treat edema / fluid retention)(ex: Lasix).

19. DMPK

(1) Drug Metabolism and Pharmacokinetics; (2) Dystrophia Myotonica Protein Kinase.

20. Don

A diminutive of the male given name Donald or Gordon.

21. Dopamine

A neurotransmitter associated with movement, attention, learning, and the brain's pleasure and reward system.

22. Dosage Form

The physical structure and appearance in which the drug to be administered is in for use.

23. Dose

The amount or form of a drug that is given to a user.

24. Dose-Effect Curve

On a graph, this is the result of plotting the dose of a drug against its effect on the bodily system.

25. Dose-Duration Curve

On a graph, this is the result of plotting the dose of a drug against its duration of time in the body.

26. Doxycycline

A tetracycline antibiotic used to treat a variety of infections.

27. Drug

A chemical used in the diagnosis, treatment, or prevention of disease. More generally, a chemical, which, in a solution of sufficient concentration, will modify the behavior of cells exposed to the solution

28. Drug abuse

The misuse of a drug under conditions considered "more destructive than constructive for society and the individual. The abuse potential of a drug depends on its capacity to induce compulsive drug-seeking behavior in the user, its capacity to induce acute and chronic toxic effects (and to permit occurrence of associated diseases), and upon social attitudes toward the drug, its use, and its effects.

29. Drug dependence

A somatic state which develops after chronic administration of certain drugs; this state is characterized by the necessity to continue administration of the drug in order to avoid the appearance of uncomfortable or dangerous (withdrawal) symptoms. Withdrawal symptoms, when they occur, may be relieved by the administration of the drug upon which the body was "dependent". Recommended as a term to be substituted for such words as "addiction" and "habituation" since it is frequently difficult to classify specific agents as being only addictive, habituating, or non-addicting or non-habituating. e.g., drug dependence of the barbiturate type.

30. Drug Dose

Amount of drug to be taken.

31. Drug receptors

Proteins present on the membrane of a cell which serve as binding sites for certain drugs.

32. Drug selectivity

The propensity of a drug to affect one receptor population in preference to others. i.e. propranolol is a non-selective beta-blocker (blocks all beta-receptors equally), whereas metoprolol is a beta1-selective blocker in that it has a greater preference (affinity) for beta1- over beta2-receptors. Selectivity is generally a desirable property in a drug as it can minimize potential side-effects i.e. potential of propranolol causing bronchospasm. Selectivity is not to be confused with "potency"; a potent drug may be non-selective or a selective drug may not be very potent..

33. Dummy

A form of treatment that is meant to have no effect on the user, yet imitates the contrasting drug in every way. This is also known as a placebo.

34. Dysarthria

Impaired articulation of speech due to disturbances of muscular control.

35. Dyskinesia
Group of involuntary movements that appear to be fragmentation of the normal smoothly controlled limb and facial movements.

36. Dyspepsia - disordered digestion
It is usually applied to pain or discomfort in the lower chest or abdomen after eating and sometimes accompanied by nausea or vomiting.

37. Dystonia
Prolonged and unintentional muscular contractions of voluntary or involuntary muscles. It most often affects the large axial muscles of the trunk and limb girdles.

Abbreviations

1.	D*	Loading Dose (q.v.)
2.	D	Dose (q.v.); also the "maintenance doses" administered after a loading dose (q.v.)
3.	d	Give, administer
4.	3 with top line horizontal (dr)	Dram
5.	DC	Discontinue
6.	D&C	Dilatation and curettage
7.	D forward slash C or D C capitalized	Discharge
8.	DEA	Drug Enforcement Administration (Regulates all controlled medications).
9.	decr.	Decreased
10.	de d. in d.	Daily
11.	deg	Degenerative
12.	derm.	Dermatology
13.	DH	Diaphragmatic hernia
14.	DI	Diabetes insipidus
15.	diag.	Diagnosis
16.	DIC	Disseminated intravascular coagulopathy
17.	dieb. alt.	On alternate days
18.	DIFF	Diferential white count
19.	dil.	Dilute, dilutes
20.	disch.	Discharge

21.	disp	Dispense
22.	div.	divide
23.	DIVA	Digital intravenous angiogram
24.	DJD	Degenerative joint disease
25.	DKA	Diabetic ketoacidosis
26.	DM	Diabetes mellitus
27.	DNR	Do not resuscitate.daunorubicin
28.	DOA	Dead on arrival
29.	DOB	Date of birth
30.	DOE	Dyspnea on exertion
31.	dr lower case	Dram
32.	D R period	Doctor, physician
33.	Drg	Drainage
34.	D5S	5% dextrose and saline
35.	DS	Double strength
36.	Dsg	Dressing
37.	D5 NS	5% dextrose in normal saline
38.	d.t.d.	Give of such doses
39.	D.T.T	Diphtheria tetanus toxoid
40.	DUB	Dysfunctional uterine bleeding
41.	DVT	Deep vein thrombosis
42.	DW	Distilled water
43.	D5W	Dextrose, 5% in water
44.	Dur. dolor.	While the pain lasts
45.	D X	Diagnosis

E

Glossary

1. **Effective**
   A situation where an administered drug is successful in attaining its purpose.

2. **Effectiveness**
   A measure of the extent to which a specific intervention, procedure, regimen, or service, when deployed in the field in routine circumstances, does what it is intended to do for a specified population.

3. **Effect modification**
   Occurs when the magnitude of the effect of a drug in causing an outcome differs according to the level of a variable other than the drug or the outcome.

4. **Efficacy**
   The ability of a medication to produce a change in its intended cell receptor.

5. **ELISA**
   Enzyme-linked immunosorbent assay, also known as an enzyme immunoassay (EIA), is a biochemical technique used mainly in immunology to detect the presence of an antibody or an antigen in a sample.

6. **Elimination Rate Constant**
   On a graph, this is the result of plotting the logarithms of concentration against time.

7. **Empiric**
Drug used to treat according to experience until a test result proves another therapy is needed. (ex: penicillin for strep throat).

8. **Enuresis**
The involuntary passing of urine.

9. **Enzyme**
Substance produced by living cells which aids in speeding up the process of chemical reactions in the body.

10. **Epidemiology**
Study of the distribution and determinants of diseases in populations.

11. **Epinephrine reversal**
Describes the response seen to epinephrine (EPI) in the presence of an alpha-blocker. The normal response to EPI alone is an increase in BP and HR. However in the presence of an alpha-blocker, EPI can now only activate the beta-receptors to cause a fall in BP with an increase in HR.

12. **Equipotent**
Being equally effective or equally able to produce the drug effect of certain strength.

13. **Equivalence**
When drugs provide identical results when administered in the same amount, or those that contain equal dosages of the same type of drug, yet are named differently.

14. **Erythemia**
Redness of the skin produced by congestion of the capillaries.

15. **Essential medicines**
Essential medicines are those that satisfy the priority health care needs of the population. They are selected with due regard to public health relevance, evidence on efficacy and safety, and comparative cost-effectiveness.

16. **Ethology**
The study of behaviour of animals in their normal environment.

17. **EudraVigilance**
The European Union data-processing network and management system,

established by the European Medicines Agency (EMA) to support the electronic exchange, management, and scientific evaluation of Individual Case Safety Reports related to all medicinal products authorised in the European Economic Area (EEA). EudraVigilance also incorporates data analysis facilities.

18. Excipients
All materials included to make a pharmaceutical formulation (e.g. a tablet) except the active drug substance(s).

19. Excretion
The drug is eliminated from the body.(usually eliminated in the urine).

20. Exocytosis
Vesicular release of transmitter i.e. NE storage vesicle migrates to and fuses with the plasma membrane to release NE (and other compounds within the vesicle ie. DBH) into the synaptic cleft. Non-exocytotic release includes the displacement of NE by amphetamine or tyramine, which can then leak across the plasma membrane in the synaptic cleft.

21. Experiment
Also called a bioassay, this is the process of establishing the strength of a chemical, physical, or biological agent, by way of a biological marker.

22. Extrapyramidal side effects (EPS)
Primarily neurological adverse events involving voluntary and involuntary musculature, including dystonias, parkinsonism, and akathisia.

23. Ex vivo
Taking place outside a living organism.

Abbreviations

1.	EC_{50}	The molar concentration of an agonist that produces 50% of the maximum possible response for that agonist.
2.	ED50	*In vitro or in vivo* dose of drug that produces 50% of its maximum response or effect.
3.	EC	Enteric coated
4.	ECG or EKG	Electrocardiogram

5.	Elix	Elixir
6.	e.m.p.	As directed
7.	et	And
8.	ER	Extended release
9.	Ext	Extract
10.	ex aq.	In water

F

Glossary

1. **Face validity**
   Judgment about validity of an instrument based on intuitive assessment of the extent to which the instrument meets a number of criteria including applicability, clarity and simplicity, likelihood of bias, comprehensiveness, and whether redundant items have been included.

2. **First-order kinetics**
   According to the law of mass action, the velocity of a chemical reaction is proportional to the product of the active masses (concentrations) of the reactants. In a monomolecular reaction, i.e., one in which only a single molecular species reacts, the velocity of the reaction is proportional to the concentration of the unreacted substance (C).

3. **First-pass effect**
   All drugs that are absorbed from the intestine enter the hepatic portal vein and pass through the liver before they are distributed systemically. Some drugs (ie. propranolol) have a high degree of removal from the circulation on their first passage through the liver.

4. **Food and Drug Administration**
   A federal organization responsible for ensuring compliance with the Food, Drug and Cosmetic Act.

5. **Fluorescence**
   A first optical system (excitation system) illuminates the sample using a specific wavelength (selected by an optical filter, or a monochromator). As a result of the illumination, the sample emits light (it fluoresces) and a

second optical system (emission system) collects the emitted light, separates it from the excitation light (using a filter or monochromator system), and measures the signal using a light detector such as a photomultiplier tube (PMT). The advantages of fluorescence detection over absorbance detection are sensitivity, as well as application range, given the wide selection of fluorescent labels available today.

6. **Fluorescence polarization**
The samples in the microplate are excited using polarized light (instead of non-polarized light in FI and TRF modes). Depending on the mobility of the fluorescent molecules found in the wells, the light emitted will either be polarized or not.

7. **Formulary**
A listing of medicinal drugs with their uses, methods of administration, available dose forms, side effects, etc, sometimes including their formulas and methods of preparation.

8. **Frequency of ADRs**
In giving an estimate of the frequency of ADRs the following standard categories are recommended:

Very common* > 10%
Common (frequent) >1% and <10%
Uncommon (infrequent) >0.1% and < 1%
Rare >0.01% and <0.1%
Very rare* <0.01%

Abbreviations

1.	F	Fahrenheit
2.	FDA	Food & Drug Administration
3.	Fe	Iron
4.	FI	Fluid
5.	f x	Fracture

G

Glossary

1. **Generic drugs**
   Drugs that have exactly the same ingredients and effectiveness as another, named drug or formulary.

2. **Gene Therapy**
   New therapeutic technologies that directly target human genes in the treatment or prevention of illness.

3. **Glaucoma**
   It is a group of eye diseases that are associated with a rise in intraocular pressure (IOP) that can cause blindness if untreated. Vision loss is caused by damage to the optic nerve. The two main types of glaucoma are open angle glaucoma (chronic, primary open angle glaucoma (POAG), and angle closure glaucoma (narrow angle).

4. **Generalisability**
   Applicability of the results to other populations.

Abbreviations

1.	G	Guage
2.	Gal	Gallon
3.	GB/gb	Gallbladder
4.	GC	Gonorrhea

5.	GCS	Glasgow coma scale
6.	GFR	Glomerular filtration rate
7.	GI/gi	Gastrointestinal
8.	GIB	Gastrointestinal bleeding
9.	Gm, g	Gram
10.	Gr	Grain
11.	grad.	Gradually
12.	grav	Gravida- number of pregnancies
13.	gtt	Drop
14.	GSW	Gunshot wound
15.	Gt	1 drop
16.	Gtt	Drops
17.	GTT	Glucose tolerance test
18.	GU/gu	Genito-urinary
19.	Gyn-gyn	Gynecology

H

Glossary

1. **Habituation**
   A psychological feeling of need for a certain drug due to its effects on the body.

2. **Half-Life**
   The time it takes for a drug concentration within the body to be reduced by one half of its original amount.

3. **Harm**
   The nature and extent of actual damage that could be caused by a drug. Not to be confused with risk.

4. **Harrison Act**
   A federal law regulating the distribution, transport, and manufacture of all narcotics.

5. **Hazard**
   A drug that has the ability to cause bodily harm.

6. **Herbal medicine**
   Includes herbs, herbal materials, herbal preparations and finished herbal products.

7. **Hit**
   A chemical compound that produces a result in a preliminary biochemical test indicating that the compound merits further study as part of a drug

discovery project.

8. **Homeopathy**
Homeopathy is a therapeutic system which works on the principle that 'like treats like'. An illness is treated with a medicine which could produce similar symptoms in a healthy person. The active ingredients are given in highly diluted form to avoid toxicity. Homeopathic remedies are virtually 100%   safe.

9. **Horner's syndrome**
It is characterized by an interruption of the sympathetic nerve pathway somewhere between its origin in the hypothalamus and the eye. The damage can either to the pre- post-ganglionic sympathetic fibers. The classic clinical findings associated with Horner's syndrome are ptosis (eyelid sagging), pupillary miosis and facial anhidrosis. Treatment: depends upon the identifying and treating the cause, in many cases there is no treatment that improves or reverses the condition.

10. **Hypersensitivity**
The necessary condition for a person to show an allergic response to a drug.

11. **Hypertension**
Abnormally high blood pressure.

12. **Hypotension**
Abnormally low blood pressure.

13. **Hypnotic**
A medication that produces an effect that causes a change in consciousness or is similar to a state of sleep.

14. **Hypothesis generating studies**
Studies that give rise to new questions about drug effects to be explored further in subsequent studies.

Abbreviations

1.	H	Hypodermic
2.	h, hr	Hour
3.	H20	Water

4.	HA	Headache
5.	HBP	High blood pressure
6.	HC	Hydrocortisone
7.	hct.	Hematocrit
8.	HEENT	Head,eyes,ears,nose, and throat
9.	Hgb./hgb	Hemoglobin
10.	H/O	History of
11.	hor. Som	at bedtime
12.	H & P	History and physical
13.	HPI	History of present illness
14.	HR	Heart rate
15.	h.s.	At bedtime
16.	HS	Hour of sleep
17.	H.T. & HTN	Hypertension .
18.	H x ./hx	History

I

Glossary

1. **Idiosyncratic Response**
   An abnormal response from a drug that is specific to the person having the response.

2. **Incidence**
   Number of new cases of an outcome which develop over a defined time period in a defined population at risk.

3. **Incidence rate**
   Measure of the frequency of the disease or outcome. The number of new cases which develop over a defined time period in a defined population at risk, divided by the number of people in that population at risk.

4. **Indication**
   It refers to the medical condition in which the drug has proven to be of therapeutic value.

5. **Indirect amine (agent)**
   Compounds that can cause displacement of NA from storage vesicles (i.e. amphetamine, tyramine). Note agents that inhibit neuronal uptake (uptake 1) can diminish the actions of indirect amines by preventing their uptake into the nerve terminal.

6. **Indirect parasympathomimetic**
   Agent that causes inhibition of acetylcholinesterase (AchE) to elevate Ach levels (i.e. organophosphates).

7. **Individual Case Safety Report (ICSR)**
   A report that contains 'information describing a suspected adverse drug reaction related to the administration of one or more medicinal products to an individual patient.

8. **Information bias**
   Error in the results of a study due to a systematic difference between study groups in the accuracy of the measurements being made of exposure or outcome.

9. **Information component (IC)**
   The Information component (IC) measures the disproportionality in the reporting of a drug- ADR pair in an ICSR database, relative to the reporting expected based on the overall reporting of the drug and the ADR. Positive IC values indicate higher reporting than expected. The IC has also been implemented on electronic health records, to detect interesting temporal relationships between drug prescriptions and medical events.

10. **Infusion Kinetics**
    The plasma concentration of a drug over a long period of time as it is proportional to the rate of the drug administration and inversely proportional to the rate of excretion and the area through which the drug is distributed.

11. **Inotrope**
    Drug or chemical that affects the contractile force of the heart.

12. **Inscription**
    Name of drug and amount of dose.(Amoxicillin 500 mg tabs).

13. **Intention-to-treat**
    A method for data analysis in a randomised controlled trial in which individual outcomes are analysed according to the group to which they were randomised, even if they never received the treatment to which they were assigned.

14. **Intrinsic sympathomimetic activity**
    Beta-blocker that has partial agonist action. Has potential to prevent bradycardia or negative inotropy in resting heart (if b_1 partial agonist) and to prevent bronchoconstraction (if b_2 partial agonist). Pindolol is prototype agent.

15. **In vitro**
Taking place in a test-tube, culture dish or elsewhere outside a living organism.

16. **In vivo**
Taking place in a living organism.

Abbreviations

1.	*i.a.*	Intra-arterial route of drug administration
2.	IC_{50}	In a functional assay, the molar concentration of an agonist or antagonist which produces 50% of its maximum possible inhibition. In a radioligand binding assay, the molar concentration of competing ligand which reduces the specific binding of a radioligand by 50%.
3.	i.c.	Intracerebral route of drug administration.
4.	i.c.v.	Intracerebroventricular route of drug administration
5.	ICU	Intensive care unit
6.	ID	Intradermal route of drug administration
7.	I & D	Incision and drainage
8.	IDDM	Insulin dependent diabetes mellitus
9.	ID50	In vitro or in vivo dose of a drug that causes 50% of the maximum possible inhibition for that drug.
10.	i.g.	Intragastric route of administration
11.	ii	Two
12.	iii	Three
13.	IM /im	Intramuscular; infectious mononucleosis
14.	incr.	Increased or increasing
15.	in d.	Daily
16.	Inf	Inferior
17.	I & O or I/O	intake and output
18.	in p. aeq.	Divide into equal parts

19.	Inj, i n j	Injection
20.	IQ	Intelligence quotient
21.	i.p.	Intraperitoneal route of drug administration
22.	i.t.	Intrathecal route of drug administration
23.	ITP	Idiopathic thrombocytopenic purpura
24.	IUD	Intrauterine device
25.	IUG	Intrauterine gestation
26.	IUGR	Intrauterine growth retardation
27.	IUP	Intrauterine pregnancy
28.	i.v.	Intravenous route of drug administration
29.	IVP	Intravenous pyelogram
30.	IVPB	Intravenous piggyback

K

Abbreviations

1.	K	Potassium
2.	k_a	The "absorption rate constant" for a drug administered by a route other than the intravenous. The rate of absorption of a drug absorbed from its site of application according to first-order kinetics. k_a is determined directly, or indirectly, as the slope of the linear relationship between the logarithm of the amount *un* absorbed and t, when natural logarithms, i.e. logarithms to the base e, are used. The half-time for absorption is computed as $0.693/k_a$, i.e. ln $2/k_a$.
3.	K_B	The equilibrium dissociation constant for a competitive antagonist: the molar concentration that would occupy 50% of the receptors at equilibrium.
4.	KCL	Potassium chloride
5.	k_{el}	The "elimination rate constant" for a drug eliminated according to the laws of first-order reaction kinetics; the slope of the plot of the logarithm of concentration against time, when natural logarithms, i.e. logarithms to the base e, are used. $t_{1/2} = 0.693/k_{el}$. $k_{el} = 2.303b$. $Cl_T = k_{el} V_d$. AUC from T_n to infinity $= C_n/K_{el}$.
6.	K_d	The dissociation constant for a radiolabeled drug determined by saturation analysis. It is the molar concentration of radioligand which, at equilibrium, occupies 50% of the receptors.
7.	Kg	Kilogram
8.	K_i	The inhibition constant for a ligand, which denotes the

		affinity of the ligand for a receptor. Measured using a radioligand competition binding assay, it is the molar concentration of the competing ligand that would occupy 50% of the receptors if no radioligand was present. It is calculated from the IC_{50} value using the Cheng-Prusoff equation.
9.	k_0	The "absorption rate constant" when rate of absorption (D/T) does not vary. k_0 describes the rate at which drug enters the body during constant-rate intravenous infusions, or during use of "sustained" release preparations for oral or transdermal drug administration.
10.	KVO	Keep vein open

L

Glossary

1. **Labeling A Drug**
    It includes the purpose or reason for using the drug and the form of the drug.

2. **Laetrile**
    A drug, derived from the amygdalin of apricot kernels, once thought capable of destroying cancer cells by the release of cyanide.

3. **Latency period**
    The period of time which must elapse between the time at which a dose of drug is applied to a biologic system and the time at which a specified pharmacologic effect is produced. In general, the latent period varies inversely with dose; the relationship between dose and latent period for a given agent is described by a time-dose or time-concentration curve.

4. **LC-MS**
    Liquid chromatography-mass spectrometry. An analytical chemistry technique that combines the physical separation capabilities of liquid chromatography (or HPLC) with the mass analysis capabilities of mass spectrometry. There are a lot of mass analyzers that can be used in LC/MS. Single Quadrupole, Triple Quadrupole, Ion Trap, TOF (time of Flight) and Quadrupole-time of flight (Q-TOF).

5. **Lead compound**
    a. A compound that has been selected from a group of hit compounds based on qualities such as the intensity of the biochemical effect that

occurs when the compound is present (efficacy), or the absence of coincidental effects (specificity)

b. a chemical compound that has pharmacological or biological activity and whose chemical structure is used as a starting point for chemical modifications in order to improve potency, selectivity, or pharmacokinetic parameters.

6. **Liniment**
A topical medical preparation intended to be rubbed into the skin with friction, such as to relieve symptoms of arthritis.

7. **Loading (priming) dose**
The first dose of a series that is larger than subsequent doses.

8. **Loperamide**
A drug effective against diarrhea resulting from gastroenteritis or inflammatory bowel disease.

9. **Luminescence**
The difference with fluorescence is that the light emitted by the samples is the result of a chemical or biochemical reaction (instead of being the result of excitation by light). Luminescence plate readers are simpler optically than fluorescence readers, as they don't require a light source, just a light detector. Typically, the optical system consists in a light-tight reading chamber, and PMT detector measuring the light emitted by the samples during the reaction. Common applications include luciferase-based gene expression assays, as well as cell viability and cytotoxicity assays based on the luminescent detection of ATP.

Abbreviations

1.	L	Liter
2.	LA	Long acting
3.	Lb	Pound
4.	LCD	Coal tar solution
5.	l i q	Liquid
6.	L R	Lactaid ringers

M

Glossary

1. Maintenance Dose
The doses in a series that follow the initial loading dose.

2. Malignant hyperthermia (MH)
It is a pharmacogenetic disease of skeletal muscle. When exposed to inhalation anesthetics (those which are gases), muscle metabolism increases with a rapid rise in body temperature which if left untreated can lead to death. Triggering agents include succinylcholine (NMJ depolarizing blocker) and volatile anesthetic. Treatment: Drug of choice is Dantrolene (inhibits Ca++ release).

3. Mean effective dose (ED50)
The dose of a drug calculated to produce a result in 50 percent of the users of whom the drug was administered.

4. Membrane-stabilizing activity (Local anesthetic action)
Beta-blocker that has the ability to decrease electrical conductance, particularly in heart (Quinidine-like effects).

5. Mean
Calculated by adding all the individual values in the group and dividing by the number of values in the group.

6. Median
Any value that divides the probability distribution of a random variable in half. For a finite population or sample, the median is the middle value of an

odd number of values (arranged in ascending order) or any value between the two middle values of an even number of values.

7. **Metameter**
A term used to label the measurement of change during biological testing.

8. **Meta-analysis**
Any systematic method that uses statistical analysis to integrate the data from a number of independent studies.

9. **Metabolism**
The breakdown of a drug into a simpler form. (usually metabolized in the liver)

10. **Metabolites**
Chemical variations of a drug within the body, a result of biotransformation.

11. **Monitoring**
The performance and analysis of routine measurements aimed at detecting changes in the environment or health status of populations.

12. **Mineral Source**
Found naturally in the earth (ex: KCl - potassium chloride; Na -sodium).

13. **Misclassification bias**
Error resulting from classifying study subjects exposed when they are unexposed, or vice versa. Alternatively, misclassification bias resulting from classifying study subjects with a specific disease outcome when they are truly not, or vice versa.

14. **Multiple Dose Regimens**: A treatment schedule for a drug in which it is given at certain intervals

Abbreviations

1.	M	Minimum
2.	m	Mix
3.	m^2 or M^2	Square meter
4.	mcg	Microgram
5.	m. et n	Morning and night
6.	mEq	Milliequivalent
7.	mg	Milligram
8.	mixt	Mixture
9.	ml, Ml	Milliliter (equivalent to cc)
10.	mm	Millimeter
11.	mOsm or mOsmol	Milliosmol
12.	MS	Morphine sulfate

N

1. **Narcotic**
   A drug that is able to create an analgesic effect, which may sometimes induce an altered state of consciousness.

2. **National Formulary**
   A reference publication produced by the American Pharmaceutical Association that gives standards of purity for each drug.

3. **Negative Control Drug or Negative Control Procedure**
   A procedure incorporated into an experiment that it should not affect the experimental system in the same way as the independent variable.

4. **Neuroleptic malignant syndrome**
   Major acute side effect of antipsychotics (neuroleptics) characterised by severe muscular rigidity, fever, an altered level of consciousness, and an autonomic instability.

5. **Neuroeffector junction**
   A specialized synapse between a nerve cell and the specific organ or tissue that it innervates.

6. **Neuromuscular Junction (NMJ)**
   The junction between the terminal of a motor neuron and a skeletal muscle fiber is called the neuromuscular junction. It is simply one kind of synapse. Nerve impulses travel down the motor neurons and cause the skeletal muscle fibers at which they terminate to contract. This is part of

the Somatic (Voluntary) Nervous System.

7. **Neurotransmitter**
   A chemical messenger that conducts a nervous impulse across a synapse.

8. **Neuron**
   A nerve cell.

9. **Non-competitive binding**
   It describes the event where there are two molecules that are able to bind an effector molecule, such as a receptor or enzyme. The molecules need not be similar with respect to chemical structure, since their binding sites to that effector are not the same. That is, the two molecules bind to different positions on their effector. Normally one of the molecules is required to activate the effector, whilst the other is an inhibitor and acts to prevent the action of the effector. Hence when considering non-competitive antagonism as an example it would mean that the antagonist does not bind to the receptors activation site and can bind either elsewhere on the receptor preventing, for example a conformation change from taking place and activating intracellular effectors, or may in fact have an effect to block some part of the downstream events from receptor activation between it and the response. So there is not competition between the molecules for binding the receptor. However, an agonist may bind to the receptor, but with no response since the antagonist is blocking the transformation to the activated state by binding elsewhere on the receptor. The distinguishing feature of non-competitive binding in comparison to competitive is that non-competitive is non-surmountable and it is not possible for a maximum response to be achieved in the presence of a non-competitive antagonist.

10. **Nonexperimental studies**
    Studies in which the investigator does not control the therapy but observes and evaluates the results of ongoing medical care. The study designs that are used are those that do not involve random allocation, namely case reports, case series, analyses of secular trends, case-control studies, and cohort studies.

11. **Non-Specific Binding**
    The proportion of radioligand that is not displaced by other competitive ligands specific for the receptor. It can be binding to other receptors or proteins, partitioning into lipids or other things.

12. **Noradrenaline (or norepinephrine)**
    A hormone closely related to adrenaline and with similar actions, secreted

by the medulla of the adrenal gland and also released as a neurotransmitter by sympathetic nerve endings.

13. Nude mouse

A laboratory mouse from a strain with a genetic mutation that causes a deteriorated or absent thymus, resulting in an inhibited immune system due to a greatly reduced number of T cells. The genetic basis of the nude mouse mutation is a disruption of the FOXN1 gene. Most strains of nude mice are slightly "leaky" and do have a few T cells, especially as they age.

Abbreviations

1.	N	The number of doses in a series; as a subscript, the last dose in a series or the number of the last dose.
2.	Na, Na+	Sodium
3.	NaCl	Sodium chloride
4.	N/A	Not applicable
5.	n.b.	Mark well
6.	NCP	Nursing care plan
7.	NEB	Nebulizer
8.	Neg	Negative
9.	Neuro	Neurology or neurological
10.	N.F.	National Formulary
11.	NFTD	Normal full term delivery
12.	NFTSD	Normal full term spontaneous delivery
13.	NG	Nasogastric , nano grams
14.	NHL	Non-hodgkins lymphoma
15.	NIDDM	Non insulin dependent diabetes mellitus
16.	NKA	No known allergies
17.	NL	Normal
18.	NMN	Not medically necessary
19.	NMT	Not more than
20.	NO_2	Nitrous oxide
21.	Noc	Night

22.	noct.	Nocturnal
23.	noct. Maneq	night and morning
24.	non rep	Do not repeat, no refills
25.	NPO, npo	Nothing by mouth
26.	NR	Do not repeat
27.	NS, N/S	Normal saline (sodium chloride, 0.9%)
28.	1/2 NS	Half-strength normal saline
29.	NSA	No significant abnormality
30.	NSD	Normal spontaneous delivery
31.	nsg.	Nursing
32.	NSR	Normal sinus rhythm
33.	NSVD	Normal spontaneous vaginal delivery
34.	NTD	Neural tube defect
35.	NTG	Nitroglycerine
36.	N&V	Nausea and vomiting

O

Glossary

1. **Observational studies**
   Studies in which the investigator does not control the therapy, but observes and evaluates the results of ongoing medical care. The study designs that are used are those that do not involve random allocation, namely case reports, case series, analyses of secular trends, case-control studies, and cohort studies.

2. **One-group, post-only study design**
   Consists of making only one observation on a single group which has already been exposed to treatment.

3. **Orthostatic (postural) hypotension**
   The gravitational stress of sudden standing normally causes pooling of blood in the venous capacitance vessels of the legs and trunk. The subsequent transient decrease in venous return and cardiac output results in reduced BP and can cause the individual to faint. Baroreceptors in the aortic arch and carotid bodies sense the change in BP and activate autonomic reflexes that rapidly normalize BP by causing a transient tachycardia and vasoconstriction in the lower limbs. Agents that interfere with this reflex response can cause orthostatic (postural) hypotension i.e. alpha-blockers, ganglionic blockers and guanethidine.

Abbreviations

1.	0	None
2.	O	Objective (soap)
3.	O_2	Oxygen
4.	O_2 cap	Oxygen capacity
5.	O_2 sat.	Oxygen saturation = SaO2
6.	O x 3	Oriented to time, place, and person
7.	OA	Osteoarthritis
8.	OB	Obstetrics
9.	Obs	Observation
10.	OCP	Oral contraceptive pills
11.	o.d.	Every day
12.	O.D.	Right eye
13.	Oh	Every hour
14.	oint	Ointment
15.	ol	oil
16.	om	Every morning
17.	omn. hor.	Every hour
18.	OM, AD	Otitis media, right ear
19.	OM, AS	Otitis media, left ear
20.	OM, AU	Otitis media, bilateral
21.	On	Every night
22.	Op	Operation
23.	Ophth.	Ophthalmology
24.	OR	Operating room
25.	Ortho	Orthopedics
26.	os	Mouth
27.	O.S.	Left eye

28.	**OTC**	Over the counter
29.	**O.U.**	Each eye, both eyes
30.	**Oz**	Ounce

P

Glossary

1. **Parameter**
   During an experiment, one of the components that can be controlled to remain constant throughout the procedure.

2. **Palliative**
   Comfort care; Treats but does not cure. (ex: Morphine for cancer).

3. **Parenteral**
   Administering a medication by injection through a route other than by alimentary canal (e.g., intramuscularly or intravenously).

4. **Parkinsonism**
   A group of neurological disorders characterised by hypokinesia, tremor, and muscular rigidity. Antipsychotic induced parkinsonism is generally characterised by the triad of resting tremor, muscular rigidity, and bradykinesia (manifested as a mask-like facial expression or reduction of accessory limb movement, or as a problem of initiating movements). Other side effects include slowed cognition, worsening of negative symptoms, shuffling gait, and excessive salivation.

5. **Pharmacodynamics**
   The study of the relationship between drug level and drug effect. It involves the study of the response of the target tissues in the body to a given concentration of drug.

6. **Pharmacoepidemiology**
   The study of the use and effects of drugs in large numbers of people.

7. **Pharmacogenetics**
   The study of the inheritance of certain interactions from drugs on the human body.

8. **Pharmacognosy**
   The study of drugs that are obtained from natural plant and animal sources.

9. **Pharmacokinetics**
   The study of absorption, distribution, and biotransformation of drugs on the body.

10. **Pharmacology**
    The study of the features and characteristics of drugs and medications.

11. **Pharmacovigilance**
    The science and activities relating to the detection, assessment, understanding and prevention of adverse effects or any other drug related problem.

12. **Pheochromocytoma**
    It is a rare tumor that arises from tissue in the adrenal gland. The tumor increases production and release of epinephrine (adrenaline) and norepinephrine (noradrenaline), which raises blood pressure and heart rate. Most pheochromocytomas are removed surgically, individuals are initially stabilized with alpha-blockers (ie. phenoxybenzamine) or alpha/beta-blockers (labetalol or carvedilol). Beta-blockers alone should never be given alone prior to administration of an alpha-blocker.

13. **Placebo (effect)**
    A medicine or preparation with no inherent pertinent pharmacologic activity which is effective only by virtue of the factor of suggestion attendant upon its administration.

14. **Plant Source**
    Bulb, root, leaves, stem or flowers: (Foxglove - Digitalis); (ex: digitoxin; quinine).

15. **Prototype drug**
    It is the 'lead agent' in a drug class (family). i.e. propranolol is the prototype of the beta-blockers and metoprolol is the prototype of the beta1-blockers. These are common agents used in exam questions.

16. **Prodrug**
It has no pharmacologic activity until converted into an active compound. i.e. alpha-methyl dopa is converted to the biologically active agent, alpha-methyl-norepinephrine (alpha2-agonist). The change may be a result of biotransformation, or may occur spontaneously, in the presence of, e.g., water, an appropriate pH, etc.

17. **Polypharmacy**
The administration of two or more drugs together.

18. **Possible drug causality**
A clinical event, including laboratory test abnormality, with a reasonable time sequence to administration of the drug, but which could also be explained by concurrent disease or other drugs or chemicals. Information on drug withdrawal may be lacking or unclear.

19. **Positive Control Drug**
A drug used in an experiment that has the expectation that its results will be similar to those of the independent variable.

20. **Possible drug causality**
A clinical event, including laboratory test abnormality, with a reasonable time sequence to administration of the drug, but which could also be explained by concurrent disease or other drugs or chemicals. Information on drug withdrawal may be lacking or unclear.

21. **Post-marketing surveillance**
The study of drug use and drug effects after marketing, which employs epidemiological methods characterised by their observational, rather than interventional, nature.

22. **Post-registration surveillance**
The study of drug use and drug effects after registration, which employs epidemiological methods characterised by their observational, rather than interventional, nature.

23. **Potency**
The strength of a drug in terms of the concentration or amount administered.

24. **Potentiation**
A special case of synergy (q.v.) in which the effect of one drug is increased by another drug that by itself has no effect.

25. Precision
The accuracy with which certain values of input can be understood by measured values of output.

26. Pre-ganglionic nerves
Nerve fibres that extend from the central nervous system to the autonomic ganglia.

27. Prevalence
The number of events in a given population at a designated time (point prevalence) or during a specified period (period prevalence).

28. Post –ganglionic nerves
Nerve fibres that extend from the autonomic ganglia to the target tissues.

29. Prevalence rate
Measure of how common a disease or outcome is. The number of existing cases in a defined population at a given point in time or over a defined time period, divided by the number of people in that population.

30. Prevalence study bias
A selection bias, which may occur in studies when prevalent cases rather than new cases of a condition are selected for a study.

31. Priming Dose
The first dose of a series that is larger than subsequent doses.

32. Probable/likely drug causality
A clinical event, including laboratory test abnormality, with a reasonable time sequence to administration of the drug, unlikely to be attributed to concurrent disease or other drugs or chemicals, and which follows a clinically reasonable response on withdrawal (dechallenge). Rechallenge information is not required to fulfill this definition.

33. Prodrug
A substance with little action that becomes more active after being in the body.

34. Prophylactic
Substance or agent used to prevent disease. (ex: birth control).

35. **Prospective case report**
Drug exposure is defined in the case prior to knowledge of outcome.

36. **Prospective study**
Study performed simultaneously with the events under study.

37. **Potentiation**
A situation where the result of one drug is increased by the use of another drug that has no effect.

38. **Protopathic bias**
Interpretating as a result of an exposure a variable that is in fact its determinant.

39. **Power** - this is the probability that a statistical test or study will detect a defined pattern in data and declare the extent of the pattern as showing statistical significance.

Abbreviations

1.	P	After
2.	pA$_2$	Measure of the potency of an antagonist. It is the negative logarithm of the molar concentration of an antagonist that would produce a 2-fold shift in the concentration response curve for an agonist.
3.	pc	After meals
4.	PCA	Patient controlled analgesia
5.	pD$_2$	The negative logarithm of the EC$_{50}$ or IC$_{50}$ value.
6.	PDR	Physicians' Desk Reference
7.	pEC$_{50}$	The negative logarithm of the EC$_{50}$ value.
8.	per	Through or by
9.	pIC$_{50}$	The negative logarithm of the IC$_{50}$ value.
10.	pK$_B$	The negative logarithm of the K$_B$ value.
11.	pK$_d$	The negative logarithm of the K$_d$ value.
12.	pK$_i$	The negative logarithm of the K$_i$ value
13.	PM, pm	Afternoon

14.	**po, PO**	By mouth, orally
15.	**post-op**	After surgery
16.	**PR**	Per rectum
17.	**pre-op**	Before surgery
18.	**PRN, prn**	Whenever necessary
19.	**Pt**	Pint, patient
20.	**p.p.a.**	Shake the bottle first
21.	**p.r.n.**	As occasion arises
22.	**PULV**	Powder

Q

Glossary

1. **Quantitative Graded) dose-effect relationships**
   Graph of the relationship between dose and response (effect) wherein all possible degrees of response between minimum detectable response and a maximum response are producible by varying the dose or drug concentration, i.e., the curve is continuous.

2. **Quantal (All-or-none; binary) dose-effect relationships**
   Relationship between dose and effect that describes the distribution of MINIMUM doses of drug required to produce a defined degree of a specific response in a population of subjects. Only two responses are allowed: Yes or No; 0 or 1. The purpose of the plot is to allow predictions about what proportion of a population of subjects will respond to given doses of the drug or toxin.

Abbreviations

1.	q2h	Every 2 hours
2.	q3h	Every 3 hours
3.	q4h	Every 4 hours
4.	Q6H	Every 6 hours
5.	Q8H	Every 8 hours
6.	Q12H	Every 12 hours

7.	Q	Every
8.	QAM	Every morning
9.	q.d.	Every day, once a day
10.	qh	Every hour
11.	QHS	Every hour of sleep
12.	q.l.	As much as desired
13.	q.i.d.	Four times a day
14.	QPM	Every evening
15.	QN	Every night
16.	QNS	Quantity not sufficient
17.	q.o.d.	Every other day
18.	q.p.	As much as you please
19.	Q(q)	Every
20.	q.q.h.	Every four hours
21.	qq. hor.	Every hour
22.	qs	Quantity sufficient
23.	qs ad	A sufficient quantity to make
24.	q.suff.	As much as suffices
25.	qt	Quart
26.	quoted	Daily

R

Glossary

1. **Random allocation**
   Assignment of subjects who are enrolled in a study into study groups in a manner determined by chance.

2. **Randomized clinical trials**
   Studies in which the investigator controls the therapy that is to be received by each participant and uses that control to allocate patients among the study groups randomly.

3. **Random error**
   Error due to chance.

4. **Rate-limiting step**
   This is slowest point in a series of reactions (i.e. uptake of choline into the nerve terminal In the synthesis of Ach) or where the enzyme involved is subject to regulatory control (i.e. Tyrosine hydroxlase involved in NA synthesis).

5. **Raynaud's syndrome**
   Condition in which small arteries, most commonly in the fingers and toes, spasm and cause the skin to turn pale or a patchy red to blue on exposure to cold or even the thought of cold. Although Raynaud's is usually a mild condition, it can have serious direct consequences, such as gangrene serious enough to warrant amputation. Treatment: Treatment: simple exercise may suffice (i.e. swinging your arms around like a windmill), however if attacks are frequent or severe, dilating agents, such as nifedipine, calcium channel blocker may be prescribed.

6. **Rebound effects**

   Discontinuation of an agent my cause exacerbation of previous symptoms to a level which is greater than before, and than that which would have been expected. i.e. sudden discontinuation of clonidine leads to rebound hypertension, tachycardia and angina

7. **Recall bias**

   Error in the results of a study due to a systematic difference between the study groups in the accuracy or completeness of their memory of their past exposures or health events outcome.

8. **Receptors**

   The part of a cell that responds to an administered drug.

9. **Reference Standard**

   A drug with specific aspects that is used as the foundation of comparison with other substances that have similar aspects.

10. **Refractory**

    Patients or conditions that do not respond to a drug.

11. **Reliability**

    The degree to which the drug and organism relationship is reproducible if it is studied again under similar conditions.

12. **Risk**

    The probability that damage will result from exposure to a specific agent.

13. **Retrospective study**

    Study conducted after the events under study.

Abbreviations

1.	R	Rectal
2.	R/	Take
3.	RL, R/L	Ringer's lactate
4.	Rx	Prescription

S

Glossary

1. **Schedule I drugs**
   No medical use; Illegal; Highly addictive; (Heroin, Marijuana, LSD) (street drugs).

2. **Schedule II drugs**
   Has medical use; High abuse; written Rx only; No refills. (Dilaudid, Morphine, Seconal, Percocet - severe pain).

3. **Schedule III drugs**
   Has medical use; Less abuse; Physical /Psychological dependency. (Tylenol #3; Didrex; Butisol; Virilon)

4. **Schedule IV drugs**
   Has medical use; Low abuse; Limited dependence. (Valium, Ativan, Phenobarbital, Talwin, Chloral Hydrate)

5. **Schedule V drugs**
   Has medical use; Low abuse; (Robitussin AC - for coughing, Lomotil - for diarrhea).

6. **Screening**
   The presumptive identification of unrecognized disease or defect by the application of tests, examinations or other procedures which can be applied rapidly. Screening is an initial examination only and positive responders require a second diagnostic examination

7. **Selectivity**
    The ability of a drug to affect one type of cell over others.

8. **Selection bias**
    Error in a study that is due to systematic differences in characteristics between those who are selected for the study and those who are not.

9. **Sensitivity**
    The ability of a specific group to respond to a drug in a certain way compared to other organisms.

10. **Septic shock**
    Serious condition that occurs when an overwhelming infection leads to low BP and low blood flow. Vital organs, such as the brain, heart, kidneys, and liver may not function properly or may fail. Treatment: Dopamine (iv) is the drug of choice.

11. **Serious reaction**
    A "serious" reaction is defined by the ICH (International Conference on Harmonization) as any untoward medical occurrence that at any dose: (i) results in death, (ii) is life-threatening, (iii) requires patient hospitalization or prolongation of existing hospitalisation, (iv) results in persistent or significant disability/incapacity, or (v) is a congenital anomaly/birth defect

12. **Side effects**
    Effects which are not desirable or are not part of a therapeutic effect; effects other than those intended. i.e. treatment of peptic ulcer with atropine, dryness of the mouth is a side effect and decreased gastric secretion is the desired drug effect. If the same drug were being used to inhibit salivation, dryness of the mouth would be the therapeutic effect and decreased gastric secretion would be a side effect.

13. **Signal**
    Reported information on a possible causal relationship between an adverse event and a drug, the relationship being unknown or incompletely documented previously. Usually more than a single report is required to generate a signal, depending upon the seriousness of the event and the quality of the information.

14. **Silent Antagonist**
    A drug that attenuates the effects of agonists or inverse agonists, producing a functional reduction in signal transduction. Effects only ligand-

dependent receptor activation and displays no intrinsic activity itself. Also known as a neutral antagonist.

15. Somatic nervous system

Controls all voluntary systems within the body with the exception of reflex arcs. This system is comprised of the afferent nerve network, which include all sensory nerves leading to the brain, and the efferent nerve network, which includes all motor nerves leading from the brain to the muscles (NMJ). The somatic system is generally associated with all body movement and is not part of the Autonomic NS (involuntary).

16. Spare Receptors

A pharmacological system has spare receptors (a receptor reserve), if an agonist can induce a maximum response when occupying less than 100% of the available receptors. The existence of spare receptors reflects a circumstance in which the maximum effect produced by an agonist is limited by some factor other than the number of activated receptors. Whether or not a system has spare receptors depends upon the nature of the receptor and its coupling to the measured response, the number of receptors, and the intrinsic activity of the agonist.

17. Specificity

The capacity of a drug to manifest only one kind of action. A drug of perfect specificity of action might increase, or decrease, a specific function of a given cell type, but it would not do both.

18. Specific Binding

The proportion of radioligand that can be displaced by competitive ligands specific for the receptor.

19. Spontaneous reporting system

System in which case reports of adverse events are voluntarily submitted from health professionals and pharmaceutical manufacturers. This system may serve as an early warning system for adverse events.

20. Standard Drug

Establishing the strength of a chemical, physical, or biological agent, by way of a biological marker.

21. Standardized Safety Margin

The amount of a drug that is effective in almost all of the population that must be surpassed in order to produce a fatal effect on a minimum amount of a population.

22. Stimulant
A drug that enhances or increases a bodily function.

23. Specificity
The ability of a drug to show only one type of result.

24. Subscription
Directions to the pharmacist - total amount to dispense and drug form. (Disp: #30 tabs)

25. Supersensitivity
An excessive amount of sensitivity to a drug.

26. Superscription
Date, Pt. name, address, DOB, Rx symbol (Jane Doe, 1234 Anywhere Dr., Ohio 44444, 10/17/1972).

27. Surveillance
Ongoing scrutiny, generally using methods distinguished by their practicability, uniformity, and rapidity, rather than by complete accuracy. Its main purpose is to detect changes in trends or distribution in order to initiate investigative or control measures.

28. Synapse
A space between nerve cells. Electrical impulses are transmitted across the synaptic cleft by neurotransmitters.

29. Synergy
The use of two drugs together provides a greater effect than the sum of the original drugs.

30. Systematic error
Error introduced into a study by its design rather than due to random variation.

Abbreviations

1.	s with line above it	Without
2.	SC	Subcutaneously
3.	sc or sq	Subcutaneous
4.	sig	Write on label
5.	si op. sit	If necessary
6.	SL	Sublingual
7.	SOB	Short of breath
8.	Sol	Solution
9.	Solv	Dissolve
10.	s.o.s	If there is need
11.	SQ, SC or subQ	Subcutaneous
12.	SR	Sustained release
13.	Ss	One-half
14.	St	Let it stand, let them stand
15.	Stat	Immediately and once only
16.	SubQ, subq	Subcutaneous
17.	Supp	Suppository
18.	Susp	Suspension
19.	Sum	Take
20.	Sx	Symptom
21.	Syr	syrup

T

Glossary

1. **Tachphylaxis**
   The building of tolerance to a drug after repeated administrations.

2. **Therapeutics**
   The science and techniques of restoring patients to health. A single drug may have two or more therapeutic effects in the same patient at the same or different times, or in different patients. Drugs may be used prophylactically to prevent disease or to diminish the severity of a disease should it occur subsequent to or during treatment; such a use of drugs is commonly called "prophylactic therapy". Drugs are sometimes used to measure bodily function and contribute toward the diagnosis of disease.

3. **Therapeutic index**
   A number that measures the relative safety of a drug.

4. **Threshold Dose**
   A dose of a drug that is just enough to produce its desired effect.

5. **Time Concentration Curve**
   On a graph, the time concentration curve is the relationship between the dose of a drug and its latency period.

6. **Time-of-flight mass spectrometry (TOFMS)**
   Time-of-flight mass spectrometry (TOFMS) is a method of mass spectrometry in which ions are accelerated by an electric field of known strength. This acceleration results in an ion having the same kinetic energy as any other ion that has the same charge. The velocity of the ion depends

on the mass-to-charge ratio. The time that it subsequently takes for the particle to reach a detector at a known distance is measured. This time will depend on the mass-to-charge ratio of the particle (heavier particles reach lower speeds). From this time and the known experimental parameters one can find the mass-to-charge ratio of the ion.

7. **Time-resolved fluorescence (TRF)**
Relies on the use of very specific fluorescent molecules, called lanthanides, that have the unusual property of emitting over long periods of time (measured is milliseconds) after excitation, when most standard fluorescent dyes (e.g. fluorescein) emit within a few nanoseconds of being excited. As a result, it is possible to excite lanthanides using a pulsed light source (Xenon flash lamp or pulsed laser for example), and measure after the excitation pulse. This results in lower measurement backgrounds than in standard FI assays.

8. **Tone (Autonomic)**
Under resting conditions most organs of the body receive a low but steady release of NA or Ach (tonic release) to modulate tissue activity. In the heart the basal release of NA contributes about +5 bpm and the release of Ach about -10 bpm to the resting heart rate. This is why beta-blockers such as propranolol can cause a fall in HR as they prevent the action of the tonic release of NA. Likewise the muscarinic antagonists, such as atropine can cause an increase in HR as it prevents the action of Ach. Usually one division of the autonomic NS dominates under resting conditions, GI-tract, eye, heart (parasympathetic) and vasculature (sympathetic).

9. **Tolerance – Tachyphylaxis**
Continual use of an agent can result in diminished response. In some cases this can appear in mins-hrs or dose to dose and is termed tachyphylaxis (i.e. amphetamines). In other cases it appears more gradual over days-months and is termed tolerance. (i.e. opioids).

10. **Toxic effects**
Responses to a drug which are harmful to the health or life of the individual. Almost by definition, toxic effects are "side effects" when diagnosis, prevention, or treatment of disease is the goal of drug administration. Toxic effects are not side-effects in the case of pesticides and chemical warfare agents. Toxic effects may be idiosyncratic or allergic in nature, may be pharmacologic side effects, or may be an extension of therapeutic effect produced by overdosage.

11. **Toxicology**
The scientific discipline concerned with understanding the mechanisms by

which chemicals produce noxious effects on living tissues or organisms; the study of the conditions (including dose) under which exposure of living systems to chemicals is hazardous.

12. Transcription / signature

The Pt's instructions; Directions that follow the Sig:, written with abbreviations. (Sig: Administer i gtt to OD BID PRN for pink eye).

13. Triple quadrupole mass spectrometer

A tandem mass spectrometer consisting of two quadrupole mass spectrometers in series, with a (non mass-resolving) radio frequency (RF) only quadrupole between them to act as a collision cell for collision-induced dissociation. The first (Q1) and third (Q3) quadrupoles serve as mass filters, whereas the middle (q2) quadrupole serves as a collision cell. This collision cell is an RF only quadrupole (non-mass filtering) using an inert gas such as Ar, He or N2 gas to provide collision-induced dissociation of a selected precursor ion that is selected in Q1. Subsequent fragments are passed through to Q3 where they may be filtered or scanned.

14. Two-state Model

A simplified model of receptor activation by agonists. The receptor is hypothesized to be in conformational equilibrium between an inactive conformation R and an active conformation R*, with the equilibrium in the absence of agonist normally favoring the inactive state. Agonists bind preferentially (i.e. with greater affinity) to the active state, and by mass action shift the conformational equilibrium such that a greater proportion of receptors are in the active R* conformation. Inverse agonists shift the conformational equilibrium such that a greater proportion of receptors are in the inactive R* conformation.

15. Type A reactions

Adverse reactions which are a result of an exaggerated but otherwise usual pharmacological effect. These tend to be common, and dose related, and predictable. They can usually be treated by reducing the dose of the drug.

16. Type B reactions

Adverse reactions which are aberrant, and may be due to hypersensitivity or immunologic reactions. These tend to be uncommon, not related to dose, and unpredictable. They usually require cessation of the drug.

17. Tyramine - MAOIs interaction

Certain foods (ie. aged cheese, red wine, figs, fermented and otherwise processed meats, fish and soy products) contain large amounts of the amino acid tyramine which can interact with MAOIs to dramatically raise

HP and HR. The tyramine induces the release of large amounts of the stored neurotransmitter, NA from the nerve terminals. The reaction, which often does not appear for several hours after taking the medication, may also include headache, nausea, vomiting, possible confusion, psychotic symptoms, seizures, stroke and coma.

Abbreviations

1.	T or?	A point in time or a time interval; frequently a time interval following administration of a drug or the time interval between doses of a drug. The definition of a specific T or ? may be explicit or may be inferred from the context in which it is found. Specific times of interest may be indicated by subscripts, e.g., T_0 is the time of drug administration; T_n is the time of administration of the nth dose in a series.
2.	$t_{1/2}$	The "half-life" of a drug; the amount of time required for the concentration of a drug in, e.g., a body fluid such as plasma, serum, or blood, to be halved. The idea of half-life is legitimately applied only to the case of a drug eliminated from body fluid according to the laws of first-order reaction kinetics. $t_{1/2} = 0.301/b = 0.693/k_{el}$, where 0.301 and 0.693 are the logarithms of 2 to the bases 10 and e, respectively.
3.	tab	Tablet
4.	tal	Such
5.	tal. Dos	Such doses
6.	tbsp, T, tbs	Tablespoon
7.	ter i. d	Three times a day
8.	tiw	Three times a week
9.	TO	Telephone order
10.	Top	Topically
11.	TPN	Total parenteral nutrition
12.	tr.	Tincture
13.	tsp, t	Teaspoon
14.	Tx	Treatment

U

Glossary

1. **Unanticipated harmful effects**
   Unwanted effects of drugs that could not have been predicted on the basis of existing knowledge.

2. **Unclassifiable Drug Causality**
   A report suggesting an adverse reaction which cannot be judged because information is insufficient or contradictory, and which cannot be supplemented or verified.

3. **Unexpected adverse reaction**
   An adverse reaction, the nature or severity of which is not consistent with domestic labelling or market authorization, or expected from characteristics for the drug.

4. **Unlikely drug causality**
   A clinical event, including laboratory test abnormality, with a temporal relationship to drug administration which makes a causal relationship improbable, and in which other drugs, chemicals, or underlying disease provide plausible explanations.

5. **Untoward Effect**
   A side effect that proves harmful to the patient

Abbreviations

1.	U	Unit
2.	u.d. or ut dict	As directed
3.	ung	Ointment
4.	URI	Upper respiratory infection
5.	USP/NF	United States Pharmacopia / National Formulary (Published every 5 years, over 3500 medicines)
6.	ut dict.	As directed
7.	UTI	Urinary tract infection

V

Glossary

1. **Validity**
   The amount of error found in the results of a scientific equation.

2. **Vasodilator**
   Medicine to dilate blood vessels / decrease blood pressure.

3. **Vermifuge**
   A drug that causes the expulsion or death of intestinal worms, such as tapeworms.

4. **Vesicant**
   Any material that causes blisters upon contact with the skin.

5. **Volume of distribution of a drug**
   The size of the "compartment" into which a drug apparently has been distributed following absorption.

Abbreviations

1.	vag	Vaginal
2.	VO	Verbal order
3.	VS	Vital signs

W

Abbreviations

1.	w/	With
2.	WBC	White blood cell count
3.	Wk	Week
4.	w/o	Without

X

Glossary

1. **XTT**

   (Sodium 3′-[1-(phenylaminocarbonyl)- 3,4-tetrazolium]-bis (4-methoxy-6-nitro) benzene sulfonic acid hydrate).Colorimetric assay for the non-radioactive quantification of cell proliferation and viability. The assay is based on the cleavage of the yellow tetrazolium salt XTT to form an orange formazan dye by metabolic active cells.

Abbreviation

1.	X	Times

Y

Abbreviation

1.	y.o	Year Old

Z

Glossary

1. **Zero-order kinetics**
   A condition in which the speed of an enzymatic reaction is independent of the strength of the substrate.

Abbreviation

1.	ZnO	Zinc oxide

Miscellaneous

1.	1 gr = ___ mg?	60 mg
2.	1 oz = ___ mL?	30 mL
3.	1 tsp = ___ mL?	5 mL
4.	1 tbs = ___ mL?	15 Ml
5.	1 oz = ___ tbs?	2 tablespoons
6.	1 g = ___ gr?	15 grains
7.	1 lb = ___ oz?	16 oz.
8.	1 cup = ___ oz?	8 ounces

References

1. Common Prescription Abbreviations
   Available at: http://www.emcp.com/college_resource_centers/listonline.php?
   GroupID=10226

2. Explore Pharmacology
   Available at: http://www.med.unc.edu/pharm/aboutus/explore_ pharmacology.pdf

3. General Pharmacology
   Available at: http://paramedicsofmanitoba.ca/storage/11-
   12%20ARML/Pharmacology.pdf

4. Glossary of Pharmacology Terms
   Available at: http://www.buylowdrugs.com/pharmacy-articles/Glossary-of-
   Pharmacology-Terms.php

5. Glossary of Terms
   Available at: http://www.who.int/medicines/areas/quality_safety/safety_
   efficacy/Annex1Glossaryof Terms.pdf

6. Glossary of Terms and Symbols Used in Pharmacology
   Available at: http://www.bumc.bu.edu/busmpm/academics/resources/
   glossary/#IntrinsicActivity

7. Introduction to Pharmacology
   Available at: http://texashste.com/documents/curriculum/pharmacology/
   introduction_to_ pharmacology.pdf

8. Introduction to Pharmacokinetics and Pharmacodynamics
   Available at: http://www.ashp.org/DocLibrary/Bookstore/P2418-Chapter1.aspx

9. Lecture Notes for Health Students: Pharmacology
   Available at: http://www.cartercenter.org/resources/pdfs/health/ephti/
   library/lecture_notes/health_ science_students/Pharmacology.pdf

10. Medword Resource: General Pharmacology Abbreviations
    Available at: http://www.medword.com/abbrevs-pharm.html#.UVrg_6DLKFg

11. Pharmacokinetics and Pharmacodynamics
    Available at: http://www.mcgrawhill.co.uk/openup/chapters/ 9780335245659.pdf

12. Pharmacology
    Available at: http://en.wikipedia.org/wiki/Pharmacology

13. Pharmacology Abbreviations
    Available at: http://quizlet.com/18463356/meas-218-pharmacology-abbreviations-2-
    flash-cards/

14. Pharmacology Abbreviations: Flashcard
    Available at: www.koofers.com/flashcards/nurs-pharmacologyabbreviati/ review

15. Pharmacology Abbreviations Word List
    Available at: http://edubakery.com/Word-Lists/Pharmacology-Abbreviation-v1-Word-List

16. Pharmacology and Clinical Pharmacology Handbook 2013

Available at: http://www.fmhs.auckland.ac.nz/sms/pharmacology/_docs/
pharmacology_handbook.pdf

17. Pharmacological Glossary
Available at: http://www.tocris.com/pharmacologicalGlossary.php#.UU8l-aDLKFg

18. Principles of Pharmacology
Available at: http://nccubiopsy.files.wordpress.com/2011/03/e5bf83e7908
6e897a5e79086e5adb8e5b0 8ee8ab96_ch01.pdf

19. What is Pharmacology
Available at: http://www.academicroom.com/topics/what-ispharmacology#.
UWLhoKDLKFg

www.ingramcontent.com/pod-product-compliance
Lightning Source LLC
Chambersburg PA
CBHW051811170526
45167CB00005B/1969